Evolution is a fraud is nothing more than a religion.

Small talk about a big subject.

By
Elder Jimmy J. Biggs Junior

Order this book online at www.trafford.com
or email orders@trafford.com

Most Trafford titles are also available at major online book retailers.

Note for Librarians: A cataloguing record for this book is available from Library
and Archives Canada at www.collectionscanada.ca/amicus/index-e.html

Printed in Victoria, BC, Canada.

ISBN: 978-1-4120-7821-4 (sc)

*We at Trafford believe that it is the responsibility of us all, as both individuals and corporations,
to make choices that are environmentally and socially sound. You, in turn, are supporting this
responsible conduct each time you purchase a Trafford book, or make use of our publishing services.
To find out how you are helping, please visit www.trafford.com/responsiblepublishing.html*

*Our mission is to efficiently provide the world's finest, most comprehensive book publishing
service, enabling every author to experience success. To find out how to publish your book, your
way, and have it available worldwide, visit us online at www.trafford.com*

Trafford rev. 06/30/09

 www.trafford.com

North America & international
toll-free: 1 888 232 4444 (USA & Canada)
phone: 250 383 6864 ♦ fax: 250 383 6804 ♦ email: info@trafford.com

The United Kingdom & Europe
phone: +44 (0)1865 487 395 ♦ local rate: 0845 230 9601
facsimile: +44 (0)1865 481 507 ♦ email: info.uk@trafford.com

Table of Contents

FORWARD

In this time of increased knowledge and prosperity there are some that have taken advantage of those who just don't know or have a full understanding of certain topics. We live in a time now of increased knowledge if you look back in time over the past 6 to 4000 years you will see that there was a slow pace of progression of knowledge and travel a kind of crawling along at a snails pace that didn't curve upwardly but ever so slightly forward at a consistent pace until for some reason we became more knowledge and traveled more. The experience of living on this planet however as of late the learning Curved has turned over the past 150 years to lean straight upward.

The Holy Bible says, "In the End of Days travel will increase and knowledge will increase."

I believe that's what we been experiencing over the last 150 years and even over the last 50 to 60 years more so an increased travel and increased knowledge. Some people will have to stop and think about the fact that maybe they are smarter than they were even 20 to 30 years ago.

When I say smart I mean the ability to do things faster at an accelerated pace much more so than our ancestors, our parents, or our grandparents. My great-grandmother did not believe that a man could walk on the moon even though she saw this with all eyes on the television screen.

This is not an unusual way of thinking because of the way of thinking in those days my great grandmother was not by herself most people believed this but the new generation brought about new knowledge and new travel from speeds of 40 miles an hour to 20,000 miles per hour so the increased knowledge and the increased travel are a fulfillment of prophecy talked about in the Bible 4 thousands years ago by Prophets, Teachers, and Preachers.

Sometimes it is sad to see learned teachers and institutions, scientists and even groups of researchers fall prey to fraudulent ideas or behaviors that are just not correct. These Are just not a part of what the true values of life are. I see this a lot today as people's minds are getting more slanted towards what they want to believe as opposed to the truth in other words people wanting to accept the 'perception of reality' and not 'reality' this is becoming more of a way life within our new world and it is also fulfilling prophecy, "for the love of many shall wax cold."

I understand how living on this planet, living in your neighborhood working on your job living in your family everyone is trying to get along and be peaceful and survive. I understand how people or trying to as an old expression that was once said, "One go will along in order to get along." This is the way we operate in society we tend to go along with situations even when they don't always totally agree with everything that we believe in some times. If it's something that's closely related to what we might like, something that we can put our hands on that's kind of close to what we want to do or believe then we tend to go along with this in order to keep the peace in our workplace, with our families, with our friends, with them, on our jobs, and within our environment. And I understand that because this is how we operate in this world, in this society, and in this family, and at his workplace, or on the street.

But sometimes a man has to take a stand and say I don't want to talk about that to my child, I don't want that spoken to my child, I don't want that told to my child, not in school, not on the streets, not in any place that I paying for, because there is something wrong with that statement, there is something wrong with that way of thinking, and there is no proof, facts, or evidence PFE in order to support that way of thinking. This is nothing more than a religion to believe in something other wise. If one can only use logical common sense, one can come up to understand that the facts just presented as they are, the truth of an issue will emerge. Now I must say this evolution has shrouded and kidnapped science evolution is not science. Evolution is an ideology about science.

I am so glad that it took me some time to write this book because between the time that it took me to write this book and between the time that this book came out the evolutionists have come up with new ideas that and these new ideas are leaping before walking, there can be no leaping before walking, one must walk first one must crawl first. Evolution is not science it is an ideology about science, science says that in a man's DNA he has a double helix however one of them is an X -and one is a Y. the female of our species however has a double X. in her reproductive DNA that is clearly marked and this tells you that this is a woman. So these are complete theories and not ideologies evolution is an ideology about science it has shrouded science but it is not a theory. What we're left with when examining evolution is the absence of PFE Proof, Facts, and Evidence and that's fine if you want to believe in a religion! Its fine one can believe in what one wants to believe in, just one thing though,

Don't call your religion science.

For my missionary mom

CHAPTER 1

Evolution is a biological impossibility

By

Elder Jimmy J. Biggs Junior

Fraud:

1 The crime of art painting money or some other benefit by deliberate deception,

2 Somebody who deliberately deceived people by imitating or imprints impersonation

3 Something that is intended to deceive people

Charles Darwin a man looking for the meaning of life a scientific way of life and that's a good thing there's nothing wrong with man's ambitions a man thinking that life as he know it just doesn't seem to be complete. This is trying to discover what do you think life should be, what would I like it to be? God gave man the ability to reach out and to seek what he thinks life should be like I think it's a part of man becoming more like God to search out and reach out we are a little 'g' little gods of course but it is still what God has put in a man soul. Charles Darwin wasn't necessarily wrong in trying to find out to research to find a way to understand the how's, and the why's we exist what have we done to exist? There has to be a reason for our existence, these are very valid and good questions and I do appreciate Mr. Darwin if you really read up on him you understand that he came to some interesting conclusions on his most important quote he said, "Noting the abundance of fossils, numerous transitional ***must*** be found in order to prove my theory." I don't know but somehow the evolutionists of today seem to run by this one statement of Mr. Darwin was a scientist and was an absolute stillness about absoluteness! He believed in what he saw and what he found or what we could place this hands on and that's understandable also but, he came to a conclusion after all of his research that if he couldn't find the missing links to his next steps then he shouldn't go beyond his personal ideas or ideology. He can't really secure the stationary places of his studies and that would be his ideologies in order for them to become theories. Man has an ideas about a lot of things and there is nothing wrong with having ideas that's good things but, to have the right ideas to backup the right theories, these are the most important things that matter and Mr. Darwin had come to this point in his life and he also had stated this continuously in the number of his writings and somehow a lot of his followers missed these important parts of his extensive research.

Just to analyze the word 'Noting' Mr. Darwin was an English subject, he was a British citizen of the United Kingdom and the word 'Noting' was a common phrase used in the United Kingdom not so much used in the American language linguists thinking. I want to define the word, noting it just means a place of stopping, or a un-passable spot, or it actually started from a knot in a tree it doesn't allow the tree to flow normally, it does not flow easily, it is a place that's a stopping point, a stopping place as of this as well but, noting came in to the English language also if you would just see a knot in a rope it does not allow the rope to flow fluidly so Mr. Darwin had come to this point that his ideas and his research was not flowing fluently to a higher point of discovery, noting an impassable point and he had said to himself unless I find evidences, missing links, something to take me past this point I am at a stopping point, I'm at a holding point, I am at a stop sign, a red light, I can not get in the checkout line to buy my groceries because I don't have enough money, and I could not get gas for my car because I don't have adequate amount of currency in order to pay for I am ordering. It reminds me of when I was living in Europe and sometimes I would be on the train and I would have United States currency and I would be in Germany and I wouldn't have the exchange rates for the currency but I would offer the American currency to the German ticket takers or ticket collectors and sometimes the German ticket collectors would just tell me that's okay go ahead because they wouldn't want to take the time to make the exchange rate from the dollar to the Deutsche Marks at the time and so I got to ride free but, at other times I would have to present the right fair or they would not allow me to get on a train. So I was at a place of stopping a place of non-forward progression and this is where we find Mr. Darwin he had come to this place where he knew that after all of his extensive studies that he was at a place of non-forward prohibition! There are a number of places that appeared to be embarrassing for Mr. Darwin and his followers that are noted a lot of these situations are totally documented and noted the question here is, why are all these talented men of science standing by Mr. Darwin? Well, Mr. Darwin's colleagues are also in need funding in order to continue with in their professions! It is remarkable to me that unless you agree with the evolutionists today within most secular colleges a great student cannot proceed forward within your career. One cannot be a direct design thinking person or the fact that we didn't just happen and that peoples origins had a direct design and reason and purpose for being here with some intent.

So what we have here is for the lack of funding and for the further of one's job, ones career, ones education, and a number of ones have gone along with this way of thinking that is contrary to the ideology of Mr. Darwin even though they have less evidence today to prove his theories than Mr. Darwin had in his day in order to prove his ideologies.

But there is one phrase that ring true in the past and it still does today, people seem to always go along with it. This is a sad state of affairs that we find ourselves in, its just sad because independent thinking should be more important to them than just going along with what ever comes along in order to get along. I remind most readers of a phrase that they should be familiar with and that phrase is, "If you tell a big enough lie, then everyone will believe it! And we know who said this! A one mister A. Hitler we need not state any of the comprehensions of the ramifications of these in locations.

Religion, noun

1, peoples beliefs and opinions concerning the existence and nature and worship of a deity or deities and define environment and the universe and human life.

2, a particular institutionalized or personal mobile system of belief and practices relating to the divine.

3, a set of strongly held beliefs values and attitudes that somebody moves by.

4, an object, like this, cause, or activity that somebody is completely devoted to or obsessed by.

5, life as a monk or a nun, especially in the Roman Catholic Church.

What am I saying here, when new light is discovered about something most people are a little skeptical about receiving an anomaly with a positive response especially when the new light seems to be totally different from the old light or the new discovery disproves the old ones. Now I can see how's and the why's of the old school not wanting to adapt to change so quickly that's logical to earth was flat to the world at one time, electricity was unthinkable as a source of energy to used in the home, and dinosaurs didn't walk with humans. But the intelligent mind has learned to receive the good things of change positively, adapt to them, grow with them, and go forward.

Otherwise we have one of the three insurgents to choose for.

1, to say that the world is not roundt; 2, electricity use is an icon in the world and 3, human beings did walk with dinosaurs! Otherwise to continue to believe that the above new light that shines on these areas and others are not true. Now one can believe what one wants to but let's call what is what it is, in order to believe in something without the presence of the proof, facts, and evidence of what one believes the remainder by default is

Religious enthusiasm.

Biology:

Biology is a exact science this is a proven theory or fact like the medical field is the study of the human anatomy as we see Dr. Peter Benton a surgical doctor on the weekly TV program ER Dr. Peter Benton is on call for any and all emergencies that may call for a surgical opinion we see Dr. Benton running into the emergency situation and shouting, **"Okay what do we have?"**

Okay yes this is a TV program but the writers of this show have employed real doctors and surgeons to coach the actors with their lines to give the show one of the real-time look of an actual emergency room in operation.

So back to the show, Dr. Benton, the doctors and nurses have received a patient, the doctors began asking questions, the nurses began taking mental notes, everything is centered on the type of patient and what is the injury.

I think the people here are some of the best in their field some of the best in the world in emergency.

The people working here have seconds to make prognosis and then, come up with the right diagnosis the professionals can come up with answer because they have been trained and they practice their craft. Now sure these people can make mistakes but normally because the biology of the human anatomy is such an exact science if they stay with in the guidelines of their skill they can come up with the right answers.

It's like math there can only be one answer.

Orlando F. Massey, Mathematician

The human body is designed specifically to do a specific thing God made it in a way to react in certain ways it cannot just happen just because we want it to happen. There are 1 million, million, million, parts to the human body and we don't know all of its working parts because of the enormity of its size but to assume that you can take all the parts of a jet plane which is normally about 1 to 200,000 parts and throw them in the air and when the parts landed back on the ground they will all be put in order by themselves and the plane will just fly as normal as if it had been put together over a period of time by a vast number of people working in conjunction to achieve a specific design.

The human body is the same we did not just come about we were designed to do specific things and we do them continually because we were designed to do them.

Here's a question case scenario for you one day you're driving along in your car on a street where you'd normally drive then all of a sudden things change, you find yourself on a table with lights flashing in-your-face people shouting terms sounds like medical terms all around you even though you don't completely understand them however, you do understand that they are trying to help you and they are trying to calm you down and help you to relax even though you are somewhat uncomfortable and disturb and the question is to yourself how did you get there?

Well, logical common sense question: if you found yourself in this position would you want a medical person over you who only had one year of high school a couple of months of medical school a few weeks of on-the-job training and had only been working in the hospital a couple of days, would you want this person working on you as you are laying on the table in the need of help?

Direct or purposely design not only applies to the medical field are there specific rules and guidelines to follow an order to complete the task of helping someone back to health if possible there are the same guidelines in almost every area of specific design or if there is something that one is trying to achieve directly. Take digging for instance if one is digging a hole trying to find a well for water, oil, minerals, or whatever, if the well is 2 feet to the right then you missed the mark! It is the same way in any area; it is the same way in anything. Now let's try to spell some words in the English language but you can only use A, the letter M, the letter T, and the letter Z, only these letters may be used. I have some words we can spell using the given letters yes words like faith, believe, hope, yes, these will work because one does not need all of the facts in order to have faith or to believe in something, however, if one wants to learn to spell the English language one will need to know the complete 26 letters of the English alphabet and they are: ABC DEF GH I JKLMN O PQRSTUVW XYZ.

This is the logic and the type of faith and scientific senses the evolutionists are insisting that we must ask people on this planet to in which to believe. You have to believe this because none of us were there then and no records, no proof, no facts, no evidence are to be found to date, in other words to actually prove something one has to have had been there. If you have not been there you cannot actually say that it actually happened to say you think it happened, to say it seems like it might have happened, this is the only way that man can actually prove that anything has happened on this planet and that is for someone who was actually been there or the documentation of someone who was there and that documentation that can be backed up by a person who was also there with a different experience with in the same experience at the same time of the person with the first experience in other words you would have to have had been there in order to prove that something is true. That you say so given the lack of proof, facts, and evidence, PFE is what I call it, 'You have nothing more than religious enthusiasm.'

To close with one last small fact the word of evolution is pronounced incorrectly there is no L. behind the E so they Americans say elvolution but the English who created the language say E-V O LUT I ON this is the correct way of pronouncing it we Americans say it incorrectly so the word is E-volution and not elvolution. To be more succinctly the letter E is a vowel and can be used in of itself so there are four syllables in the word and not three so the word is pronounced differently and correctly by the English who first pinned it.

Chapter 2   Butter cups

Bob Ballard says there was a flood.

Over the past 20 years.

I found that I became intrigued with the flood when I was a kid I remember memorizing this album of Bill Cosby's, 'The Right Album', the first half talked about Noah and the Ark you know as a kid I was taking this stuff pretty literal because I had heard about the Ark in Sunday school pretty much like everybody else had so I thought that this was a good reference and hey, it was Bill Cosby I knew that he had know? He was an adult and I thought that everybody went to Sunday school. Sorry, I was a kid. I thought Sunday was named Sunday because it was always sunny; the funny thing is how as an adult now looking back this was some serious stuff to me.

On Bill's album the lord comes to Noah and says that he wants to Noah to build an ark Bill goes on to mimic the Lord with instructions as to the size of the Ark by this weight and by that size in this length and this cubit and that cubit.

As the Lord is finishing giving this information to Noah, he replies, "Lord?" Noah asked, "What's an Ark?" The Lord's Answer is, "How long can you tread water?"

Life seems so crazy sometimes but we go on. In 1975 I was with a group of friends and they were talking about this film they had come across this film was about a man climbing Mount Ararat. Well, I didn't know what a Ararat was? This film was kind of amazing.

It was in black and white I didn't like black and white films so I had started to lose interest with black-and-white movies because they were before my time. Now don't get me wrong the people really had to act in those days because they didn't have the graphics and the computers, special effects if you would, and to compare the two it's not the same on the scale the actors back then had to much more disciplined, most of the time they had to do it in one take not given all day to or a week hind of lie Soap actors.

I'm spoiled by color and graphics like most people I like a lot of action and adventure movies, different things that keep my attention because otherwise I would probably be going to sleep. It's the society we live in now days this is what people are trying to text message, to twitter while they are driving. It's really fast nowadays what we know is that we are moving really faster when completing today things. Life has really sped up and people are getting a little bored if things are not going fast or if they don't keep the attention span because they are very short now days. Anyway back to 1975, in the film of these people the film was about a guy who went to Mountains of Ararat with the crew of people from France, I really lost interest but I tell you what I regain interest and I cover this in the next chapter 3 with Dr. Bob Ballard he is the man that proved the great flood did exist. For how to look for the great flood, this is something that was supposed to have happened over 4000 years ago. How do you look for evidence, of different types of water residue? Now this can take some time because where do you start? Well Dr. Bob looked for places of different types of water, there are different types of water of course we all know seawater, salt water, lake water, ocean water, ocean water with seawater, freshwater, and there are still many more different types of water. Dr. Ballard for some reason went to the Black Sea to look for different types of seashells or different experimentations I really don't know and it would be hard for me to speculate because this particular sea is underneath the mountains of Ararat in fact right down the mountain or the place of the most sightings of Noah's Ark. The findings of Dr. Ballard for some reason found seashells at the bottom of a salt water ocean however, these were freshwater seashells.

This is highly unlikely this is something that one would never find because the two different types of water do not cross in fact he would be hard pressed to find these two different types of oceans with similar types of organic life. So the two are normally never near each other Dr. Bob Ballard did a complete study on this and found just what people said couldn't happen.

There are some on this planet that wouldn't want anybody to know that evidence on the great flood is available and not just hearsay. But if you look at the facts you might find that there is proof of the great flood in lots of places one only has two get by the brain damage of what the popular scientists are saying and read the newspaper for one's own self.

You have to get past the brain damage that you have been taught in school because there are more going on in them there hills oh yes more than it is thought to have been believed.

Butter cups in the stomach

In most situations when talking to scientists on how the fossils were found at higher locations and at mountain regions and talking specifically about the fine of how does buttercup flowers found in the stomach of an baby elephant not digested when the animal was found frozen and intact.

This goes right across the grain of the ideology of the evolutionists a baby elephant found with undigested buttercup flowers in its stomach is a large of questions and answers however, not the answers that the evolutionists want to surmise.

Before we get started I like to ask a question: How do you make of fossil?

If an animal just dies on the grass it doesn't become a fossil. If you bury an animal he doesn't become a fossil. If an animal or a human for that matter falls into a glacier or a large body of ice and water and freezes this also does not make a fossil.

The way to make a fossils is first the body must remained alive, then drowned, and then packed over with tons of mud and earth and water over a long period of time with the fluids, skin and full muscles intact this preserved the bones so that as the skin, muscles, and all internal parts as they are deteriorating over the course of time they act as a buffer so that the earth would not scatter or shatter the bones. This is the key issue in the making of the perfect fossil this is why the fossils found over the last 150 years are perfect and on file today. So there is no other way to make a fossil except using the results of the great flood.

If an elephant had undigested buttercup flowers in its stomach yet it was found frozen and intact the question is how? It could not have had survived all of these years and it could not have evolved. An assumed year of birth and the fear of death was assigned to it but nothing can ring true to the evolutionists idea of where a mammal of this type should be, more less how did it become frozen in time?

See the elephant had to be frozen, or fallen into a glacier, or flash flooded, but the evolutionists say that there was no flood. Because if there was a flood then there was a Noah and if there was a Noah then there is a God that still lives and reigns and created everything that exists in all of 400 billion stars of existence and this seems too large to comprehend especially for one super being. Man's problem is he believes that he is the center of the universe and all of life revolves around him. Now given the fact that this is the only planet in all of the universe that we can perceive that has life, makes man think that he is the center of it all and that he can't fake his way through anything because we have the intelligence regardless of the fact that we only use one 10% of our brains capacity but man's ego is 100% of a man's ability and 0% of what men can achieve when following this edict.

In fact let's go back how do you make a fossil well, if an animal dies on the street doesn't turn into a fossil? I don't think so.

If the animal is caught and frozen to death do you think that this will become a fossil? I don't think so.

In order to make a fossil, the fossil dies, we receive it intact, and it takes pressure and water and tons of mud impacting on animals or being in order to preserve the spectacle. To be frozen won't necessarily do it because the proper flood will drown the subject thus allowing most of the bones and internal parts such as lungs, heart, etc… to stay in tact.

Only a flood will create the proper fossil because the pressure from the mountains of water and of mud and earth can exact proper death in order to preserve the fossil thus.

So the only way to make a fossil is with the flood they could not have been frozen.

So again we see from taking time in order to look at just the facts of the matter we come to the right and proper result that leads us as researchers to discover that this also had a specific design.

The fish migration from the land to the sea

The theory of evolution goes like this the fish of the ocean overtime started to migrate to land from the sea.

Logical commonsense question:

What kind of fish, which ones? Over the course of time for we(humans) came to the land and this is how we get to the next step in evolution as it goes from swimming fish to walking breathing air fish.

Logical commonsense facts:

The true facts are that there is no (PFE) Proof, Facts, or Evidence, to support this missing link.

Logical commonsense question:

Why would a fish come to land? All the fish needs to survive is water, food, other fish. Where is it written that fish were joined together with all other fish of all time to move out of water for the existence of fish everywhere or for the existence of all life? These are not decisions that fish make at least we have not seen these type of decisions being made by fish in our lifetime neither over the last 2 to 4000 years.

Logical commonsense and fact:

In the same way a computer can be delivered anywhere on this planet given the cooperation that we humans have instituted. Humans engage, illustrate, organize, and cooperate, to that and many other joint ventures. In the animal kingdom the animals do not cross the line to work for the analysis of anything as a group on any large-scale post, present, and no evidence of future. All creatures live to survive and we do what we have to in order to survive we do not perform things in order not to survive it is one of our first basic rules and instincts for living. A man might lie, steal, some may take life, in order to survive but to throw oneself in harms way for no good reason such as is the claim by the evolutionists as the fish coming to shore if a person is doing this to save the another's life or if the choice is between the lesser of two evils, we may condone this as a last resort to help someone to save someone in other words and man throwing himself in front of a speeding bus is not trying to survive in fact this man is trying not to survive. This makes me think about the heroes aboard the flight that crashed outside of Pittsburgh Pennsylvania on 911 I think most Americans and people in the world know and understand that these people had two choices I think they decided to go thier way and not the way of the terrorist. It seems that they decided to go there away which still ended up being a bad choice but the lesser of the two choices so in this way they went out their way as one man said, "let's roll" and a group of brave souls did. If the question is ever asked what a hero is, here's an answer even so, this selfless act of these brave souls and the one that they chose they chose also as survival for others if not for themselves.

These did make this type of decisions but, the theory of evolution or the evolutionary scientists goes on to say that some of the fish came to land and some of the fish died but some of the fish made it to land lived.

Logical commonsense question:

I can hear my cousin saying now some of them well which fish which ones the stronger fish live to survive for the school this seems logical yes! My cousin yells, "Well who says that you are stronger than me?" And he would go on to tell you and show you how he's stronger than you.

One biological fact:

If some of the fish died, then all of the fish died why? It is a biological impossibility to develop lungs from Gills.

This has never happened, there is no **PFE, Proof, Facts, or Evidence** that this ever has happened, that this is presently happening, or that this will ever happen in the future. So this is one of those things that one would have to believe, and have faith in, because there are no working models, no clinical studies, and no examples.

Sounds like a religious experience, religion is okay!

Logical commonsense question:

If the strongest fish in one school came along and destroyed the strongest fish in a small school and the weakfish within the small school would now be the strong fish in the school then the weakfish would be the strong fish but the school would be weak so how could the school survive with the weakfish? Given the evolutionist's ideal,' Survival of the fittest.' That's the evolutionist's ideology but that doesn't always work out because what happened if all the weakfish died then we would not have that type of fish at Red lobster? Thank God, hallelujah.

The evolutionists also say that the cow returned to the water from land in other words this is how we got the warm blooded animals, mammals, the seals, the Dolphins, the killer whales, they all came back to the sea from land, the Black Angus and the hosting Cows migrated to sea over the course of time some died the stronger ones lived of course.

This reminds me of a **logical commonsense question:** a small boy asked me some time ago, this question, 'How does a brown cow eat green grass and give out white milk?'

It was an interesting question to be sure but, I could explain it to him scientifically thank God.

Logical commonsense question:

Why would a cow go to sea? To put on scuba gear, in order to see the little fishes? All that a cow needs is own land, grass, air, other Cows! Why would cow go to the sea, the oceans? And seeing other cows die why would other cows continuously walk in the water and drown themselves? Because if the cows keep drowning themselves in another million years or so and a couple of them start swimming in the water and on the water and develop blowholes and water lungs, because we(cows) have to repopulate and make cows in the water because this is the push and the drive that we all are receiving for dolphins and sea mammals to be born for forever from our insides, we (cows) came out of the water now we are returning to the water, it is just in us to do these things! I'm sorry cows don't do these type of things they don't make these type of decisions. It sounds very unlikely for cows? I think so!

Hippopotamuses with their thick skin's stay in water in order to keep cool but they don't live in water, however, the animals nature is one of migration the Hippopotamuses do not have water breathing lungs like seals they remained in water in order to keep cool because their skin is very sensitive and they can hold their breath for long periods of time but they're not water animals.

What a biological fact: if some of the cows died then all of the cows died it is a biological impossibility for animals to develop water lungs, blowholes, and gills from air breathing lungs. Even with any other water breathing biological apparatus.

This has never happened there is no **PFE Proof, Facts, or evidence,** that this ever has happened, there's no **PFE** that this is happening now, there's no **PFE** that this ever will happen. This is one of those things that one will have to believe there in, have faith in, or have hope of, or maybe this just didn't happen? Sounds like a religious experience? Hey, religion is okay! One can believe what one wants to this is not a crime.

Each of us as human beings do not feel that push from all life in the universe and the cosmos saying to us, 'Move forward push! We must move all of existence to the next level of evolution!' Sorry, I don't think so! This sounds more like Hollywood saying, "Use the force Luke!" And if this is that indication, then that can only be preceded by default by faith!

Evolution is a biological impossibility.

CHAPTER 3   NOAH'S ARK.COM

We can all go on now and die!

It's so hard sometimes to understand how some people could accept the general populace attitudes and statements and will adhere to and just go on with their lives as if what was said is true without asking the,

Logical commonsense question:

How can you go forward like mindless sheep without asking the question without really knowing the answer, without asking about or asking how did you come up with the answer? In school sometimes you have to explain that and sometimes you don't,

But how can you go on when you really don't understand something that is as

Important as this, how life began or the origins of man, or where do we come from, where did it all start, is there a God, who am I, what is my relationship to the rest of life, what is my relationship to the rest of man, woman, Indian chief? These are very important questions and they should be answered with very important answers.

To quote Bishop Marshall Carter his response to young people when they thought or think that they know all the answers after living 80 plus years of his life for the Bishops has gone on to be with God, would say to young people, "Patients, most young people think they have all the answers? However, he was known to retort to them, "You can have all the answers because you haven't heard all of the questions yet." This is not just a statement this is a true statement. Life is full of unanswered questions life is full of questions on answered life. Life is full of sometimes, you have to wait, you have to experience, you have to look for the answer and sometimes the answer still won't come until years later. However, some questions have been answered, But to just accept the ideology of evolution just because the evolution scientists said so without any examination, explanation, research, and to just acceptance is absurd, risky, and irresponsible even for a five year old child.

Wild Chicken!

From 1972 to 1976 I was in the United States Air Force I am a Vietnam era veteran that means that I was in the United States military during the time of its engagement in that country. In the short of the military, I was engaged I did my time, and I was let go with the honor.

However in 1975 I was stationed in a place named Wasskuppe, Germany I must say my time in the armed forces was pretty good all things considered my first duty assignment was at Key West Florida at a place called Boca Chica Island so it wasn't too tough a duty I guess I had it good. It was considered as a remote site and I volunteered for it in order to be close to home and got it the first time I applied. I remember landing there at the small airport flying in on a six seat plane landing at this very small airport I think no one was there in the complete airport on the day I arrived I have to call the base in order to get someone to come pick me up because I didn't know where to go I was at the airport but where do I go from here, no one gave me any instructions. All things considered the assignment was a lot of fun, for a year and a half.

I lived there and I had lots of fun I met a lot of God-fearing men and women and I also became God-fearing in Key West with some people that love God.

I did leave home one way and came home to another my missionary mother was there to receive me and it was a good thing. Key West was very nice warm breezes, blue waters, the people still wait at the end of the dock to watch the sun go down every day it's a very nice, it's a very nice setting it is one of the most southern points of the United States. However after a year and a half believe it or not I was getting tired of paradise and I wanted to see the world and the world was provided at the age of 21 I receive orders on my birthday to travel to Wasskuppe Germany.

The airbase there was very small only 100 people just a very small close-knit group of people. this was a place in Germany were the first German fighter pilot plane crashed in World War II. We had to have a support base and one was provided a large Army Post it was called Wild Flicken. Now these are German names in German town's very nice places but it was said about Wild Flicking that one could get anything that they wanted anytime of day or night in any amount, size, shape, form, or fashion it was that of kind of place, this is the reason the troops dubbed it Wild Chicken.

Fernand Navarra

This is also where I met Lieutenant Colonel Peter Craig United States Army Corps of Engineers Chaplin. Lieutenant Colonel Craig was always trying to entertain us guys there on this large post. The post included a Chapel there and it Christian soldiers,. This is where I saw a film by a Mr. Ferdinand Navarra this is the film that was in black-and-white that I didn't show much interest to because I thought it was an old movie but after I start looking into the movie it started to become very interesting this gentleman and his crew had traveled to Mount Ararat twice once in 1955 and a second time in 1969 a lot of people didn't believe him the first time so he filmed the second excursion with about 10 plus people climbing up the mountain. One side seem to be a little harder than the other to climb the closer you get to the top he said it seems as if the mountain was retracting back underneath your feet you take one step up and slide one step back. I want to point out right here that there was at least 100 different sightings over the last 300 years of a boat like object on the mountains of Ararat in exact same place from people all over the world, the United States CIA have detailed files with pictures from satellites from infrared pictures, steel camera pictures taken from airplanes and helicopters, that include a boat like object that is still based on the mountains of Ararat and they can be viewed from the archives upon request. Also people from around the world who have also visited the site and taken pictures the United States military and other countries, civilians, and military personnel alike, from airplanes and helicopters, and from just people hiking to the mound. Also I like to add from the film that I saw they were talking about how the history of most of the people on this planet presently starts from the villages and towns down to heel from the mountains of Ararat. The Zuckerman brothers also have found different writings that from Noah himself that gives details of how his three sons and their three wives, the children of Noah's Ark went down the mountain but Noah stayed on the Mountains of Ararat praising and thanking God.

So getting back to Mr. Navarra after his second trip he return with documents and filming events that took place during the expedition's climb. This was included on film the expedition talking about they would take the other way around this time because of the surface of the mountains as a very unstable slope on the mountains.

Sometimes one can be a little bit narcissistic and I thought that everyone knew about Noah's Ark and it was no big deal yes? I everyone knew, no. I thought everyone knew the story of Noah's Ark so I didn't think it was a big deal, not knowing that this was one of the only films ever made of an expedition to mountains of Ararat.

Thank you, Chaplin Craig.

As the film proceeded forward I could see Mr. Navarra being lowered into the Ark with a rope with the help of the men with him on the expedition into frozen ice I could see him coming out I could see his coworkers bringing pieces of wood out with them.

Mr. Navarra talked about the place looking like a boat fixture and a large place with a number of stalls he couldn't travel very far into the structure because of snow and ice damage and denigration of the actual structure. After a while after collecting a number of pieces of wood as much as they could carry the crew and himself Mr. Navarra and his son began to be seen going down the mountain being very happy this time at least he had some evidence of visiting the mountain he could finally proved that Noah's Ark was on top of Mountains of Ararat. Now the whole world could share with him, Mr. Navarra's joy and enlightenment of the wood that was brought back. All of the wood the crew brought back down the mountain was tested the crew was overwhelmed the wood was cruel gopher wood (gopher wood doesn't grow on this planet anymore) so this means 1, this was an actual wood from Noah's Ark or 2, if this wasn't Noah's Ark what was and how did this man-made boat looking structure arrive on the mountains of Ararat at the 14,000 foot level, a gopher wood-based looking boat that fit the well-known description and location in the well-known areas as described as Noah's Ark. of the Bible?

This documentary went on to describe and talk about the people in the towns down the hill from the mountains of Ararat as they reached the expeditious return the family went on also to show how some of the people had made different artifacts from the gopher wood. The word was filed publicly and documented. The film was made available to anyone that wanted to view it. I didn't think much of its at the time I thought that everybody knew about this I always thought that I was one of the last people to know about these type things come to find out again, only a few people even viewed this film only few people ever heard of Mr. Navarra, Mr. Ferdinand Navarra..

In my own personal research of history of the world I found that there were many roots of history in the area of the world the mountains of Ararat the rest of the world could find some of its historical beginnings, and it is reasonably so there has been a lot of research conducted in around the mountains of Ararat.

As viewed in the documentary film in the 1980s, 'In search of Noah's Ark', there is a segment illustrating the search for the Ark of Noah that also talked about sightings from a number of people through the years but there was one other site and among other places that talked about sightings of a large object but in two different signing that were observed. When they were discovered that the ship had broken into two places one at the 17,000 foot level and one part at the 14,000 foot level. This could be only view from space because the closer you get to the top of the mountain the harder it is to breathe and to survive temperatures would also become very cool so in order to view the site very clear it could only be done from the satellites and these are a part of record these are part of history documented and very well known. And I say again along beside myself a number of people have either seen or reported sightings of Noah's Ark from almost 300 years prior and there is a great number of people listed on the Internet from countries all over the world from all walks of life, all over the world, from total civilian, total military, and almost everything in between. Noah's Ark has been cited, and I have been saying that Noah's Ark has been discovered. I have one question for National Geographic. Why haven't you engaged a gigantic search for Noah's Ark? Do you know that the mountains of Ararat are the only place that National Geographic have not visited in great detail as they have completely detail the whole world, map the complete earth, discovered almost every single thing that crawls, creeps, and sleeps on this planet, every single detail of every single minute thing from the jungles of Africa, South America, Antarctica, the North Pole, every possible culture every possible thing that people eat on this planet from snails to snakes, from snakes trying to eat people. Yes, lions, the Tigers, the Bears, oh my! National Geographic's has visited,

However, they have never ever excavated the mountains of Ararat in full out search of Noah's Ark in great detail as they have excavated the rest of the world. Science has explored large parts of outer space as far-reaching as the eye can see National Geographic's has also traveled to as far deep in the world's oceans as man can travel 2 to 3 miles down I think it's a 7 mile down trip to the end but as far down as man can go National Geographic's has been there but, yet feet 14,000 up to 17,000 feet, National Geographic's can't seem to make it there why is that? I challenge National Geographic's 14,000 feet to 17,000 feet? But the veracity they have excavated a small part of the world to see what they will find.

For years as Noah's Ark rests on the border of the Russian, Turkey and neither country would allow anyone to visit the mountain for a long period of time. As a question, National Geographic's, why haven't you?

If anyone can discover Noah's Ark at the mountains of Ararat that means that there was a flood and sense there was a flood that means that there was a Noah and that means that if all of the above is true then there is a God. But other religions believe in God so it shouldn't matter right? Even the Jews and Arabs are from the same father they are half-brothers they are of sons of Abraham and so those religions believe in God right? Wrong!

How can this be wrong? Noah's God is not everybody's God. See Noah's God is the same God that destroyed the world and everybody's God do not destroy the world, everybody else's God Allows you to live like you want to live. To do whatever you want to do. Noah's God said he destroyed the past world because of their evil ways Noah's God said that's why I put you in the boat because I am destroying the world because an their evil ways. Noah's God said I'm coming again with Jesus Christ and everyone needs to be ready because everyone will be talking to ME. Everyone will kneel down and give praise and humble themselves and give honor to me because I created the world and everything in it. This is Noah's God. It took Noah 120 years to build the ark and there was no rain at all on the planet as we know it. Noah built the giant boat in the middle of nowhere and no one believed him everybody thought he was crazy but I can imagine no one thought this especially after the animals came inside the Ark. It was thousands, maybe tens of thousands. The Ark is huge now the animals just came to the Ark he didn't have to go and chase any, looked for any, all of the animals came to buy in order at the exact time, at the exact amount, the exact weight. The rains started coming down especially after all the animals were inside as God closed door to the Ark and sealed in Noah, his wife, his three sons, and there wives and all the animals inside the Ark. I can imagine Noah looking out and thinking to himself about the people as the waters began to rise and the people begin to beat on the sides of the boat of Noah the same ones who had picked at Noah for over 100 years, the people Noah and his family daily tried to in courage to join him and his family before the rains stared and after they continued to ask people to come with them. And yet there were none that answered their plea. Sadly,

The only words that come to my mind for hours and maybe even days are the words of Bill Cosby,

"How long can you tread water?"

Evolution is a historical impossibility

CHAPTER 4   NOAH'S ARK.COM

In search of Adam and Eve and the genome code

Knowledgeable

When we find that we are not knowledgeable and unlearned or just don't totally understand something about some subjects it is understandable because we can learn anything with the proper research and the proper tools when we have so many available avenues from which to resource. Libraries of the world were always within our neighborhoods and now with the increase access of the internet there is almost nothing one can't study or research or area where we can't find answers to almost anything that we are interested in learning, can't we, or not?

As we move forward in our new world we are discovering all kinds of things that give us new meanings to areas we previously thought otherwise was of no effect.

People really did think that the world wasn't round because they couldn't see past the horizon everything looked flat people couldn't comprehend sailing around something they could only comprehend what depravity they had. Remember there was no satellites built, no Internet, no telephones, there was no photographer, no telegraph, there were none of those things, there was just men and boats and ships and horses and carriages they did really travel very far and apparently the people didn't travel around the world. And as far as the sun rise and the sunset most people did see this change of stars in the sky only the change of moon and the Sun. Which did constantly change however, the people didn't think about the earth revolving around this huge Sun they just thought it was day and then it was night the sun went up and the sun went down. The number of hours in the day time was a constant number of hours of my time, but most were not thinking that we are all revolving around in space. All they wanted was equal amounts that change the seasons this made a lot of things that man has learned that has changed through the years in time and man totally. Differently that he did not have knowledge about those things that are totally changed from the way that they are now. Even if you look at electricity in the house do you know people actually were afraid to install electricity in their home? People thought 'we will all be electrocuted inside of our houses if we put electricity inside'. Of course thinking about their houses to electrify them, to make light, to use electricity in any way, what would happen if a lightning storm occur and these are valid pre-cautions because those are the things that could happen in the lightning storm to a house today if not properly ground. And it's remarkable how people thought this way in those days and people still think this way somewhat today. In those same ways but somehow we have gone past this way of thinking and we think a little differently now we see the things that man won't acknowledge and that is this is an increased knowledge. It is an increased ability these are things that have happened to us in these last 150 to 200 years. Our knowledge has surpassed that of the Roman Empire which was of its day the greatest way of living at the time. The Romans Empire made some great strides in architecture and a great number of disciplines. However, they still had nothing when compare to what we experienced. None, even the Egyptians who lived 4000 years ago they had a great new ideas, in mathematics, even in Aviation and Aeronautics. Most of the cultures that rose out of Egypt had great minds even though, they still had no true comprehension of what we have in comparison by today's standards. Be that as it is the first Chinese shogun warrior (and I'll talk about this later) was from Africa. And that being an African is not so important, what question is important is, how did he get there from Africa to China? He sailed yes, Africans did a lot of sailing they had great sailing ships back in those days but today he we do more than sailing today we don't think about putting electricity anywhere and in anything we teach our children to use it continuously anywhere while we've made it safe for all so we can use it almost anywhere in any thing but has knowledge, increased has traveled increase? I say yes, because they're interconnecting with our brothers and sisters around the world we've learned to do this with a little bit more validly and their knowledge also has increased within our knowledge of reading and joining because all of us were building a tower of Babel as a monument to man's achievement together before our language was changed at the Tower of Babel this is a historical event not just a biblical tale if you can research history it'll take you back to a time when all people spoke the same language so I've made some advances, we've gotten some knowledge, and we've changed as well. Covenant People have common knowledge that Ben Franklin and George Washington walked around the house with candle operas in their hands at night to see their way through. There were no modern devices to help them see the light at night.

As fast as a man could travel, was as fast as a horse could take us. So from 40 miles an hour to 20,000 miles per hour as the space shuttle flies in less than 200 years..

In search of Adam and Eve

September 1988

Newsweek magazine, '**In search of Adam and Eve**' was the heading interestingly enough to my surprise I say to my surprise because I do expect a lot of truth from the evolutionary scientists nonetheless more missed information or the truth bent to their advantage is what has emerged as of late. Or maybe they were just trying to sell magazines. On the front cover of the magazine was a color portrait of two African descendents a male and female in the background was one of an artist depiction of the garden of the Eden.

This had to be looked into, as a researcher of history myself and one who looked for the meaning of life from an early age I always tried to understand the why's and how's. I was sure that the magazine and newspapers were only out to sale the articles and out to sale papers that's what they do, that's business and I understand but this article seem to have some teeth besides I trusted Newsweek as being a reliable source through the years from my first reading back in the 60s.

When I saw the two people on the magazine cover I thought that this backs up some ideas or ideologies that a lot of people have had through the years and the idea of the first people on the planet being from Africa because of the genome or the genes which I will cover later and it makes a little bit more sense giving the more dominant genes of the African genes and the more recessive genes that come after that form different gene pools.

As I started to read the latest fine by the science community that was on display as far as the genome was concerned this was to give an answer to where did we all start? Displacement was discovered somewhere just north of South Africa there is an area that is considered to be the oldest relics of where people lived in the world. It is a cave like place where human tools were discovered you know, mom, and dad, the kids, there was sleeping, bugs bunny and the like(of course not bugs bunny I added that part) at any rate, this is the oldest living evidence to date that has been discovered on our planet.

Bryant Gumbel of the today show (Bryant is working for another network at this time) and the rest of the crew at NBC was the first morning America show and morning talk show, news program, or television program to broadcast live from Africa and they did this for seven days.

I remember Bryant stating that he had been working on going to Africa for five years so this had to be a great accomplishment for him and being an African descendent the rest is clear. The show made number of stops throughout Africa with in a seven day tour. I remember seeing Bryant on the now off the air television show The late-Night Talk show with Arsenio Hall a couple of days before Bryant's trip. Bryant Gumbel after being introduced or announced by Arsenio Hall was to be seen walking up to one of the persons in the audience that had said that they didn't like Bryant and Bryant was shaking the man's hand and saying something like, "I'm sorry that you don't like me." Don't quote me on this but it was pretty close to something of that nature. Bryant after taking his seat next to Arsenio Hall went on to talk about this subsequent trip and the broadcast from South Africa coming the next week and one of the stops the show visited was at this historical place. I remember that the area was still so fragile that today no one is allowed to walk on the site or touch any of the articles on the site for fear of disintegration so people can only view it now a days from 20 to 30 feet away as in times past people were allowed to actually walk on the site of the oldest living place where people sat and ate and it appeared to be a homeowner landing. It is still there just north of South Africa today. From this area the scientist concluded research and things were interesting to tell a few things.

The Genome code

The genome code encompasses the DNA the building blocks of our existence. This tells us what type of person we are through the genes that are aligned with them. It's kind of like a ladder that turns and twists through ever cell of our bodies mapping the code tells us what part of the world we are from and where we get our genes. So now back to the Newsweek article of September 1988 and what the article and the scientists have found or discovered and surmised was that all life as we know it started in Africa this is shown in the dominant genes in the makeup of every human on the planet. The most dominant gene in every human being on this planet is an African gene and this gene also came from an African female. This is well documented and shows that everyone on this planet was born from an African female. This is the same gene that is the most dominant gene in every human being on this planet and in their DNA.

So it doesn't matter what you look like on the outside or how you look on the outside, or from where you were born in your dominant gene in every human being that has ever lived and that ever will live on this planet came from an African woman so therefore everyone has a little African-ness in them one way or another. No one on this planet is further than 50th cousin because of the genetic makeup that is in every human being that has ever lived on this planet. However, the apes and the animals are not included in our genetics.

This also is the reason that our anatomical parts can be interchangeable more often than not. This is also explains why they're all the same in other words to the mouth, lungs and hearts, livers and the like are interchangeable and but they are not the exact same, no two are alike. Sure we do not have to have a perfect match but when we get close the matches are close enough with help in order to interchange. That's because there are no two of us that are alike. This tells you that it's impossible and that there is no way that we could have evolved. Because if you look the situation if no two of us are alike then how could any one of us ever become more than one of us? If it was solely dependent upon the next person in front of us being just like us because the next person behind us wasn't like us and we were not like them and the ones in front of us are not like us even down to your ear. Do you know that every ear is different and distinct; there are no two ears on any one person or any person at all on this planet history of human being that are alike? 'To each, his own'. This thing is about the fact that not even identical twins, the ones that are so chose the ones that look so much alike are not totally identical because there are no such thing as two of the same thing on this planet given that in this situation presently. Announcing this, we could not have evolved because to move all of man onto the next phase within the same stage and personalities are terribly different, from a lot of other things in a larger range of things, not to mention what eat, don't be ready to sleep with the lights on, look at what you like to wear.

It is interesting to me that one can take the same shoe and give them to 100 different people and you will come up with 100 different ways for the shoe to be worn The people can live on the same side of town with in the same village within an area of only 1000 people. However, they will all come up with a different way to do this shoe. Individuality eradicates evolution. So since a small group of people within the same town, with in the same village, within the same city of only 1000 people and yet you will receive 100 different ways of doing one thing. How could tens of thousands of people evolved if two of them could not agree on anything? Was this a **logical commonsense question?**

This is our nature but what if you just individualize every shoe how would one complete anything repetitive. With the ideal of evolution everyone would have to be the same? I don't think so! Everyone is totally different and as the evolutionists say that in order to progress from one stage to the next there has to be repetitiveness there has to be somewhere in the stage where everyone is moving on to the next stage because they are all alike at the biological microscopic DNA genetic place? Sorry, we all are not the same at the biological microscopic DNA place! Because you see that's where the changes take place not on the surface with her hair color or eye color or his skin color, that's not where changes take place. Changes in anatomy takes place at the microscopic level at the atomic level where the atoms, neutrons, and electrons are, this is somewhere in the middle of your DNA that's where all if any change would take place. So if no two people are alike from the ear canal, to the eye, and no two noses are alike how could we have evolved to anything? No, no, I say to the contrary we did not evolve to anything we were designed to specifically to something! Designed to be something this is academia, intelligence, and logical commonsense. Yes, that, but not just be what ever would come about if left alone to route all devices and that's what evolution sounds like just come alone no matter what and there will be Law and Order to control all the parts or to the bill a jet plane and throw the parts in the air and when they land on the ground it all comes together as one fully functional jet plane that we can now **go on board** into an take off to Florida? Give me a break please?

Speaking of blood types here humans have different blood types 0 positive, B negative, a list of different types of blood and some are interchangeable in order to operate on persons in a emergency. But animals anatomy are not interchangeable with humans. Animals have one blood type and their blood type it is not interchangeable with humans. Animals have one blood type humans can't use animal blood.

There are a few different revelations that come along with this ideology, is this the consistency of how humans haphazardly came about by random selection? This is not responsible and it's not reasonable, rational or factual given that no of us are alike but that we are like? One other thing in this area we are similar we have the same lungs, hearts, ears, nose, the rules are the same except they go a little further, and I reiterate no one on this planet is further than 50th cousin nobody on this planet no human being on the planet. This only applies to human beings; there is a big leap to the animal world. The percentages seem small between humans and animals however, the blurred lines are very long and thick and different, and wide. So this shows that we humans are all descendents from the Africans, and then the Asians, and then the Europeans. As we humans are more scientifically named the Negroid and the Mongoloid and the Caucasoid. This is the order of all humans the genes are more dominant in the Negroid then the Mongoloid then the Caucasoid. The reverse is also true recessive gene starts from the Caucasoid to the Mongoloid back to the Negroid these are scientific facts, complete theories and documented Proof, facts, and evidence present. This is not racial at the lab since this is not an ideology, is not someone's thought or idea these are scientific proven realities. As humans moved out of Africa and started toward and to lived in cooler climates their skin color became lighter for lack of more exposure to the sun, their hair became thinner, they retain less protein in their hair, so it is less curly and because of the lesser yearly amount of sunlight it became lighter.

Cracking the code

So let's move forward and two 2001 cracking the code of the genome. The scientists had all begun to understand the code back in the 80s the guys and girls back in the lab were making progress of cracking the code in those days however, there was so much that the group didn't know. But, there were a large number of things that were known and that they did learn that they didn't know at the beginning. In later days things begin to look pretty amazing, ideas, hope, and promise to those that worked for a number of years.

When the code was cracked what the scientists received was a parts list in other words they new all of the parts of the genome now. This was an amazing discovery because maybe now it could be more identification, interaction, experimentation, elimination. The list took the ideas of scientist off the chart. However, all things take time. One scientist explains and says genes have extra 'stuff' he no other retort or explanation for this exclamation other than he could see with the genes how it made up in the code with 50 to 60% of the total code was double extra and it was duplex and duplex genes. One thing I found amazing was that less than 10% of the DNA code was given to skin color, hair color, and type, and color of eyes. We place much emphasis on how we look, where we look, the color of hair, skin, eyes to fashion, a buyer would place much emphasis on those things yet genetically they are the least among all other parts of the human anatomy. Well just think of it as genetic make up that your skin even though the epidermal is the largest organ on your body, your skin but it is the least of all dominance when it comes to genetics within your body's organs when compared to leg, muscles, capillaries, arteries, the spleen, liver, ventricle aorta, the list just goes on and on and on of the different parts of the body that are so much more important than the hair color, eye color, if we place more emphasis on those things then the most important things then it wouldn't matter what color you are without a heart you have no blood pumping out to the rest of your body, without kidneys you would not have fluids clean throughout the rest of your body it does not matter what color your body is on the outside the inside of all bodies are pink, red, some veins are green or blue but those are the same colors inside of the body regardless if the exterior use from the darkest brown to light pink they all are includes within our lives and they each are colors all combined with in that mix.

X. and Y. and XX

Even more so amazing about the code is the female genes they seem derived from the male genes. Now that's my understanding and assumption, I come to this conclusion based on one fact and I could be proven scientifically incorrect but since there are only two of these particular type helix in this case you may understand why I apply it in the genome code of the male. There is an X. and a Y. helix this tells you that this is the male this is the male's code. If assigned to a technician to examine different types of blood one trained to examine under their microscope the results will read only one of two types. This type of helix X. and Y. will tell you that this is the male's blood of this code origin of that type. This is what I'm talking about when it comes to evolution, to go further if this was a female's blood that was obtained it would have a double X. helix this would tell you this is a female(Means 'fetus carrier' in Latin). These has been scientifically proven for a number of years and are not secret, this is science. The reason I say the females genes seem to derive from the male's genes is that he has both genes and the female does not. The similar gene is the X gene this is the same gene within both beings, this is why I surmise my ideology.

Asexual procreation is linear only to asexual reproduction

I've never received an answer to this question from evolutionists and I will pose this question also to you;

Evolution is based on one cell developing into one fish, them into one lizard, then one mammal, then one monkey, then to one human. (I left out a few stages for this particular statement here so you bear with me please) this had to have been a male because only the male carries both the male and female DNA the X. and Y. chromosome and the female only carries the double X. for linear reproduction. Given that the people on this planet are linear procreative males and females.

Logical commonsense question:

From where **did** the female come? Why did I choose the female? Simple, there are two types of people on this planet a male and a female, the female cannot reproduce herself unless she was the asexual female that means she was born pregnant giving birth to children that were also born pregnant and they also gave birth to children that will be born pregnant and that will go on and on and on. However, we are not all asexual females born pregnant giving birth to other asexual females giving birth and on and on and on. So then this must've been an asexual man, right? Born pregnant giving birth to a asexual male and him giving birth asexually to asexual children that are born pregnant and on and on as the same ideology in the movie Godzilla? Where this giant lizard was born a male and do to some genetic interaction with a nuclear blast became an asexual male and laid 200 eggs and they also became asexual males at birth. Now this is an interesting idea and it is probable but, is not possible because it has never happened like a lot of the theories and ideologies by evolutionists. But the facts are we are still not all asexual males or lizards there had to be another partner two beings procreating in order to produce one person. In other words we are not asexual males or females and so how does the evolution surmise to beings from one where there is no PFE, no proof, no fact, and no evidence, to support their ideology? There is no working model to measure from in order to derive ones synopsis. Therefore it is not a theory and yet an ideology. It is an ideology and not a theory at all; it is an unproven and incomplete theory and does not emerge to be considered by the scientists, technicians, or institutes of higher learning.

This is why I say that it appears that the females derived their DNA from the male because in other words from where else would they have gotten it

Logical, common sense, scenario:

All humans on this planet were born for one female and all humans that has ever lived on this planet at any given time this includes all humans that are on this planet at this present time or that will be born from one female that is a fact, that is a proven scientific fact, that is true. The first human to walk on this planet was a male. Why? Because I stated above only the male carries is the X. and Y. helix and the female only carries a double helix in her DNA chromosome strand. This fact means that the female couldn't have birth both male and females at best she could birth only females, even though are we all females? Not! So we come to the conclusion or we could apply Occam's razor (Occam's razor says, given all things are equal, the simplest answer tends to be the right one.)

This is the commonly used and best abbreviation of what Occam was saying he was a man from Germany. However, what he was basically saying is, if you didn't add anything to the solution or experiment or reality and didn't take anything away from the solution, then the simplest answer with in the solution tends to be the right one.

This adage tends to apply to almost every analogy, experiments, and rationale that science has ever applied it to. There are exceptions of course but I think Occam's razor applies here.

Asexual procreation is linear only to asexual reproduction in other words we can only get asexual from asexual and all beings produced from asexual can only reproduce asexual beings!

Logical, common since question;

How or where did the female come?

A sexual procreation is linear only to a sexual reproduction

I know I said that twice I don't like to be redundant but in this case I want to be clear these are facts presented in evidence not by me alone but by countless, scientists, technicians, and people research years before I was born. This had been completely documented and filed long before I made my statements. So I am basically repeating someone else's handiwork and this is nothing brand-new because however, I want it to be remembered and not forgotten but it is well known and I am probably the last person to receive knowledge publicly or state it.

The evolutionists believe this but have no evidence supporting the splitting cell ideology; They have no working model to reference from there by it is incomplete and not factual or true. This is how they come up with two beings to separate beings having procreation linear procreation to offspring to children to on and on to procreation however all evidence to the contrary there is no proof, no facts, no evidence, PFE in order to support the evolutionists ideology therefore this is another one of their incomplete theories. The first person to walk on this planet was a male and that's because God designed a man and No evidence to the contrary also indicates that God made female from male how God did that? I have no idea. The Bible says that God opened up man's side took out one of his ribs then close the flesh there after and made a female and present her to the man and he said this is bone of my bone flesh of my flesh and he loved her and took her to be his helpmeet and the two became one flesh.

They produced all the children and all DNA that man has researched, documented and filed. This shows that all humans were born from this one female and there is no evidence to the contrary. No proof, no facts, no evidence, no PFE to the contrary.

We have learned through scientific experimentation that a female can only carry a child or an offspring because the female body is stronger internally and the males body is strongly externally the word female means fetus carrier and any female organs will be needed in order to carry a fetus. So anyone claiming to be a male yet having a fetus has to retain the female internal linear reproductive organs in order to reproduce a human. So therefore this human is still a female regardless of their desire to be considered as a male, males do not reproduce, males do not carry fetuses, males do not get pregnant, only females can or are allow to and can deliver what is called the offspring of our species and the male does not have the internal capacity or ability to do so it is a biological impossibility for a human male to linear reproduce a child. The facts are if a female's overy could be interjected into a man's system the male's testosterone within his normal biological system would attack and destroy the fertilized egg. Females genes are also unique to the female body, bone density is higher, female temperature is higher the female need less water than the male. There are a lot of differences between the male and female other than it seems on the surface just wanting something is different at the microscopic level. There is so much to consider most of the time it would take years for most human beings to document the time it would take in order to attempt a reproduction on any other level within the presently accommodation.

One may dream of the idea of this fantasy but that is all that is a fantasy because in reality is males have more testosterone, females have more estrogen, males fertilized eggs the females give birth, they become children or off springs these are true factual scientifically proven statements.

Also looking at the female body isn't as strong externally also the strongest female is not stronger than the strongest male on our planet but the female athlete tends to suffer a little bit more as she builds up more of the testosterone within her natural feminine body. Her chemistry tends to reject her eternal strengths,

In short, very few top female athletes can reproduce after developing their bodies to a point of championship status there is always an exception to the rule but females on the whole at a certain level their internal workings are not made to be pushed pass a certain limit. This also excludes reproductive ness. Now all females do not reproduce sometimes this is difficulty in a small amount of cases but most females still have the capacity to reproduce probably by some medical assistance given that they do not develop their bodies past the point of no return. In order to compete and win the gold medal status it is a tossup or for most females they win the gold and the world however, they lose the children these are facts and they are un-disputed. This is not my personal speculations this is a scientific PFE proofs, facts, and evidence that are present with in medical science from years pass.

So in conclusion: In order to answer the question, the egg is the byproduct of the linear procreation and reproduction of the chicken and not the chicken being the byproduct of the linear reproduction of the egg.

In other words if you don't have two chicken you can' t. linear reproduce a chicken.

There is something I'd like to add to this world of genetics,

Dr. Chang

Dr. Chang was the physician to the Emperor of China almost 3000 years ago he had to come up with the way to practice medicine on his Emperor, that was the good news, the bad news was that the good Dr. Chang had to practice his craft without touching the Emperor! Talk about a challenge? How would he come up with this? On earth how could he do this?

How did Dr. Chang work this out because remember 3000 years ago there were no computers, no calculators, no cell phones, and no electricity inside rubber insulation!

What did Dr. Chang come up with because he didn't really have a choice of saying no to practicing his craft on the Emperor? To refuse would not have been a good choice for life to the doctor some 3000 years ago in China. So Dr. Chang put on his thinking cap and discovered Acupuncture this is a discovery of over 3000 years ago of over 3000 acupuncture puncture points that circumference the complete human anatomy. While Dr. Chang discovered this he also classified them into 12 medians and eight sub medians that are connected to every major organ and artery throughout the complete body,

This was no accidental, random selection, survival of the fittist, manufacture of the human anatomy. This was a direct design, specific of One intelligence and intentional Mind and Dr. Chang discovered it over 3000 years ago.

Dr. Chang discovered that the body has its own natural healing medicines when the body is out of sync or out of Yen and Yang or out of balance the body will be sick. The body can heal itself when it is in balance or in Yen and Yang. This is also a great part of the wonderful workings of Chinese medicine.

Dr. Chang was the first to understand and to treat anyone. So you see Dr. Chang served his Emperor very well and lived long

This practice is still very highly used today all over the world and in a number of health Care service systems in different countries around the world acupuncture can be used as a health and medical care alternative in a large number of cases people use it instead of modern medicine it is very widely used. The modern world of medical had translated the Acupuncture Science of medicine from Chinese for over 800 years.

What does this have to do with evolution?

The evolution is claimed that the body came into being totally by accident on natural selection we have shown that the body is too complex of an organism to have just come in to play by accident there is nothing in this life that we do that happens by accident the accidents are not accidents they can be explained Everything in and around us takes a plan overt or covert, some plans need a lot more preparation and explanation in understanding maybe even illustration and others situations all we at the present observation may not know the plan of the accident however, nothing happens without a plan causation.

Dr. Chang understood this also the father of acupuncture, acupressure and Chinese medicine.

Evolution is Anthological and Genenemologcal impossible.

CHAPTER 5 HUMAN FOOTPRINTS IN STONE WITH DINOSAURS

Some 30 years ago now I think I also remember viewing another film presented by Colonel Peter Craig. He was a good friend at the infamous Wild Chicken (wild flicking, Germany).

This film was talking about some people working near Glen Rose Texas at the Pulexy Gulf. I had visited San Antonio Texas but I had no idea about Glen Rose Texas of course that was further south near the Gulf of Mexico.

I remember the film talking about the fine of human footprints and dinosaur footprints side-by-side walking together, again to me this just didn't seem to be very noteworthy it just wasn't something that was interesting to me because it seemed to me if I'm the last one to find out about this then everyone knows it. And at the age of 20 years old this is just not at the top of your want to get to know meter! It is just not a part of your vernacular. Even thought, I stayed and watched the total film and found it to have some very interesting discoveries, however, this still didn't have as much important information as I would have liked. But looking back it too, this film this was one of one of only two or three reproductions of people talking about the find some 30 years ago of dinosaur footprints and human footprints side-by-side carved in stone. Lots of people making a very big noise about it, well, what's the big noise about the big noise is the evolutionists say that the dinosaurs came before the humans and the humans and the dinosaurs did not exist together and yet we have proof, we have facts, we have evidence, PFE, that dinosaurs and humans it coexists, did exist together. So given these findings this gives questions to the evolutionist's ideology of how all life began of his origins because obviously something is wrong with this picture and I don't want something other than the truth about this picture being taught to my child in school because scientists will tell you what they want you to know sometimes as oppose to what is true. Scientist will tell you the truth and the truth is if you take a drive, flight, or walk, to Glen Rose Texas at the Pulexy Gulf you also may view the dinosaur tracks side-by-side with the human foot print tracks. Dinosaur footprints tracks and humans for print tracks walking together in a number of areas and not just one or two steps or places but in a large area as if they were even walking together and there's a lot of other artifacts and facts to be found down in those areas from a number of researchers I can't necessarily mention because I don't have their permission. However, there are a large number of researchers and institutions that surround the developed, and have been excavating the area for at least 30 years. But if this is true, then there was a flood, if there was a flood, then there was a ARK, since there was a ARK, then there was a Noah's Ark, and since there was a Noah's Ark, there is THE God.

This is a problem to be evolutionary they don't want to believe that one super being created all of this it just seems too impossible but I say to you my God does the impossible, he is the Supreme being, there is no other beside Him. He created universe and controls everything in it. There is no other god, He is worthy and all will bow down to give praise and honor to Him for He is worthy of all praise, and worship, honor, and all majesty, for Jesus Christ is the son of God and God the Father is the Creator of all life and His Holy Spirit indwells all those that believe in him and those that have eyes to see and ears to hear and understandings of the things I have been writing. And that many, many, more before me have been writing and talking about, prophesying about, and teaching about the true facts that man was designed to be a specific thing to live the specific way and that way was to know the God that made him that created him to worship that God to understand the universe as God made it with limited, right here, right now, in this world. But the world cannot comprehend what God had created for us coming down here putting on the flesh of men living among us dying for us rising again the third day just for our sins so that we might live in eternity with the rest of creation. Of course God has many, many more creatures out there but he won't allow us to interact with them because they obviously have not sinned and we would not be allowed to interact with them until our sins have been cleaned and washed through the blood of the lamb which is Jesus Christ.

I'd ask scientists this question why is it that we can't get to any other living planet that we've dialogue? Since questioned why is it no other planet life can come to us?

If there were other people on the other planet that could get to us they would! If they could come, they would not hide in a farmhouse or in the middle of the desert.

They would come downtown New York City, Washington DC and talk to President Barack Obama, Michelle and the two little girls. They would talk to everyone and everybody they would not shy away to some farm field with a few farm animals and elderly people alone, they would see them and talk and get to understand more than a handful of people. Many more people would know about it for the rest of the world to see. We have a date with destiny but not yet.

If it looks like a duck, if it talks like a duck? If it walks and leaves the foot prints in stone side-by-side next to human footprints, then it's a dinosaur!

Let's get back to the dinosaurs as I looked at the films I remember seeing men taking cast of the footprints in stone and examining the stone, examining the cast, dating the cast dating the Flintstones to see if they were a fake, a fraud, to see if someone had made them recently. They search the world for a man's foot to match the stone in a file once a man who was almost 8 feet tall his foot matched almost exactly or close enough to the human footprint. It was what they were looking for but this man had never been to the Pulexy Gulf at Glen Rose Texas so it was not his footprint but it was a foot that could fit in a print like his. This gentleman was found in Minnesota I also thought the same thing of this film as I assume about the Noah's Ark expedition film. I thought I was one of the last people to see this film. Little did I know that this was one of the earliest moving pictures at the time of the documentations on the subject?

Sometimes if you really want to get to the end of the Coke bottle you have to turn it all the way up. Just turning it level to your face won't always necessarily empty it. You have to dig deep to really find answers; one cannot go on face value having going along without asking questions, questions about anything that's no surprise. That's not a big deal, that's not an ancient to the man, says the young man. I don't believe in God, proved it to me. You may not be able to prove it, however, you can do research to try to find answers to your questions because answers are out there and more answers than one would thinks one has to look long enough for. I think we live in the world where a lot of people want answers at their doorstep. You may have to travel 11,000 miles to find your answer, it may cost you money, it may cost you nothing, you may still have to make the trip just to find out and see looking for funny things that as though sometimes you have traveled 11,000 miles to find an answer to come back home and find out that the answers are waiting for you in your own backyard. If you had looked you may not even needed taken the 11,000 mile trip! Yet, I'm sure he had lots of fun on the trip, made lots of friends they created great lasting friendships and it was all great but you may not have had to take the trip at all but he didn't take the trip at all. Sometimes I've ever found out that all the answers that you looking for most times are right there underneath your nose.

Outside of the Glen Rose Texas in a place called Pulxey.

For over 25 years researchers of archaeology have been drawn to camp and have moved to and now live in this area. It is an interesting find something new to deny, something that's true only because it's different, something that shines the light on something. For instance to say that a jet engine is not faster and more efficient than a traditional propeller engine begs the phrase, "Denial is not just a river in Egypt."

The smart fellow will go up to the point of understanding at all costs solely based on common sense alone, the logical common sense person.

Here's a scenario, you have your life all wrapped up into your career, your home, or family things are set your happy. then someone comes along and tells you that the way that you were going is incorrect what you have been studying for years is the wrong way of thinking some have gone out of their way in order not to see necessarily but to increase the understandings of what they desire and of what I will accept as truth. And most secular colleges if you are not an evolutionist then you won't proceed very far when it comes to a career in most sciences, the origins of man or in any area that has any research on the findings of dinosaurs or human artifacts that deviates from the status quo. I don't necessarily fault the instructors or the institutions however, I do find it not understandable that most of them are not God fearing and are atheists. There cannot be a middle ground this seems un-compromise to the cause of just looking for the truth, oh I'm sorry that's philosophy and my other major. Philosophy always looks for truth for PFE.

Brain damage

The last 150 years evolutionist have always come up with one more way to prove that we have existed for so many years and what they normally do is use different types of testing carbon dating, and the other types of dating material. The way they do this testing is they apply to carbon dating or any other method with these instructions. 'How long have we been here one to five million years give or take 500,000 years? When in fact the carbon dating says yes thereabouts, They might as well ask the carbon dating if we've been here for two minutes because that's how long it's taken them to come up with the idea to figure out how to carbon-based something, how the big bang began because the figures are always wrong they always add too many zeros when I talk about the numbers used in dating the universe in years I used only thousands of years and never more than 10 to 12,000 years on any given situation are given and the evolutionist are always talking about billions of years but it could never be billions of years of existence as far as the universe is concern there is only evidence of beening here for thousands of years. Since we're all created together we cannot espouse on the universe past. All of these speculations are wrong because we can only prove what we can see, because no one living here now was there then tens of thousands of years ago.

One other thing that's most important one cannot say that they were anywhere if they can't prove it or to rephrase you can't prove anything unless you were there so who was on this planet 1000 years ago that is here today nobody but God. No human being was here and since no human being was here, there is no way to tell anybody that you can prove that you know exactly what happened. One can guess all day, one can do so all day but, there's no way completely prove what happened in a certain situation by a certain situation at a certain situation unless one was actually there and then it has to be crossed reference by someone of a different experience within the same experience in order to prove that it actually happened in other words take the Holocaust a way that you know the Holocaust happened is because the German guard was there and a Jewish prisoner was there. These are **two people** sharing the same experience of having different experiences with in the same experience the German guard beating to the Jewish prisoner at the same time in the same place experiencing the same things have a different experiences. The two of them can verify everything that happened in and around the town, the village, their family, they can give closure to their witnesses to the ground to the air, their site, they're seeing, if feeling, all of their five senses because the two of them were actually there sharing the same experience, sharing different experiences with in the same experience that's how you know it really happened because as to all the rest you can not have one person giving a testimony alone they can't say I know two people. They can say or add something but not more than what two people and like experiences. But two people of different experiences can so you take a person who has had one other person that had the same experience that you have had and then add 10 to that number and it's completely true and there are a lot more than 10 persons in World War II but, when it comes to evolution there is not one person today living that was living 1000 years ago and more so not tens of thousands of years ago so what do we have to verify what can we believe? The only things we can use is our things that are written down. Thing as a car, and stolen things that are made in stone so we have written documentation up to the 4000 years old we have fossils footprints and we have man's footprints and dinosaur footprints side-by-side. So what does this tell you? If it walks like a duck, talks like a duck, walks side-by-side next to humans in stone then it's a dinosaur walking side-by-side in the Pulexy Gulf near Glen Rose Texas.

As we dig a little deeper into the fines in the Rose Texas as they were initially discovered on earth the first to dig there were evolutionist scientists and they believed that they were not true and even that they might be forgeries this is the reason for the experimentation finding peoples for that match dating the footprints and making sure they were not a fraud or forgery but after a long time of research and a long time of actually examining and excavating the area more things were found, fossilized hammer, fossilized finger, that when x-rays showed the different joints of the finger. There was an actual human being these also dated at the same time as a dinosaur footprints and the human footprints walking side-by-side in the middle of the walkway. On the hammer that was un-earth people of the 21st century can not remanufacture it, cannot be duplicated the metal urge and the time placement are different from years past when compared to today in our generation in 21st century, it cannot be reproduced. The hammer could've only been reproduced a few thousand years ago but only a few thousand years ago not a few million years ago this is where the truth comes and the evolution wants to tell you that these things must've happened millions of years ago to adding too many zeros it could've been millions of years ago and lasted as long as it is even the fossil records all the fossils if they have existed millions of years ago they too would've disappeared there would not be a dinosaur fossils for us to discover.

Evolutionist try to say that we've lived here for so many tens of millions of years given the amount of people that have lived here and a world war is always taken large numbers of people away. People pass away every year and then it has been like that all of our existence but the problem is this if this planet had been here as long as evolutionist say we've been here then the amount of people that would've lived on this planet that has lived on this planet would be something like 10 to 89th power to understand this imagine what this means for a moment. If there are not enough planets in our solar system to hold that many people so that many people could not have lived on this small planet for that many years and a small planet would not still be here. There isn't enough planets in the universe to house that many people they could've lived on this planet for that length of time, it is just mathematically impossible there is just no place to put anyone that could've lived on this planet for that length of time. Not as the evolutionist would want you to believe 10 to the 89th power that's more people than we can actually think about. So basically it's impossible for the planet to have been here for the length of time than the evolutionist wants us to believe. Because there haven't been that many wars on this planet to kill off the amount of people that they say should have lived on this planet so since there have been wars disease or famine then the people would still be here and they would be much older. We would be living with them but that many people couldn't be on this planet for instance there is 6 billion people on this planet now and that has just come up to 6 billion people over the last 6000 years that wasn't 6 billion people here 2000 years ago neither 1000 years ago you come up to 6 billion with in the last thousand years the last few hundred years so if you just multiply that times a few billion years as evolutionist will want you to think that's got the multiply pass the jillions of people way pass the place the population of the planet. So, where did the people go if they could have gone even if they had died? The amount of graves and ashes that they would be buried on the surface of the earth would be stacked to hundreds of miles into the sky into the atmosphere and the stratosphere because that's just how many people it would be, we would have been living on top of hundreds and hundreds of miles of gravesites on every inch of the planet if we were here to survive we would've been raised in gravesites in burying people on the planets on land or sea almost to the atmosphere, stratosphere into air so we could not even see. This is just a biological impossibility to add that many people to live on this planet for that length of time for us to still be here and them not. We would not need a spaceship to fly from planet to planet we could just walk even out of the universe that's how many people would have lived here according to the evolutionist incomplete, unproven, ideology. You have to unlearn what you have learned.

You have to research completely what people say it goes to the magnitude of their discoveries because it appears that most react the same way to the something at the same time, this is one of those things. Also this planet would not have been here for that long a time with that many people living on it, could it? This planet could not survive if the people would've survived. It would have been too many people it just could not have happened it is just not possible. Do the math yourself calculated just for the amount of people, the amount of time, the amount of years old, one small planet, if there had only been a small group people passed away because of the wars and then of natural causes? It has been this way on this planet and even with them in the last few hundred years people still would have multiply past the point of occupation. So given that there still would be more people on this planet than let's say times eight thousand years, do the math because we have limits.

So we have another impossibility by the evolutionist.

But the final verdict is conclusive two different footprints were made at the same time and space, two different creatures are identified by man a number of sets of human species foot prints walking side-by-side and a number of sets of dinosaur species foot prints walking side-by-side in adjacent to each other at the same time in space, side by side with the same place at the same time in life. The hammer, the finger, and other artifacts, were also found at the site these can also be dated as the same time an era as the dinosaur and human footprints

Some might think that I am little hard on the evolutionist but it is far from the truth for at one time in my life I also thought that maybe the evolutionist are right? But then I took evolution out to dinner and it acted real badly in the restaurant so good that we can't go back there again. Evolution acted up so that I and evolution can never eat together again and all because I just wanted to know who evolution really was. Maybe I shouldn't have asked that question, that **logical commonsense question.**

The fine is authentic

This has been cleared for 20 years so why is it that the science community had not acknowledged this excavation as it is real right and all right. We'll get back to that some places on the later version. With a lot to lose life, for the love of money, reputation, limitation of lives, were present career status would have a lot to lose can you see the whole of science losing their jobs, tenure, the support their families? Where being incorrect about something like this would show up stock market, probably crash this world exchange? What happen? Science the same as a child would when light has been shown on their secret? How would they take care of the situation? They would cover and deny make up a big story that goes something like this, I think that this is too big of a science for the common man to understand! The layman would leave this part of science to people that seem to know best. But they don't know more than the layman when it comes to this fine city. Sometimes one does not need a surgeon to remove a splinter from one's own finger some things are logical common sense truths. This also goes to fraud. Knowing that something isn't right for proceeding however, going along with the boys and the girls for a little more money, it is? If you tell a big enough lie or deny everything then everyone who believes in anything will believe it.

Almost everyone but not everybody thank God!

Logical, common sense:

If you have such a strong belief in something that you're willing to lie or denied knowing that the truth is out there. When you think so strongly about something you believe so strongly I think this could be classified under religious enthusiasm!

Enthusiasm noun
1. passionate interest in or eagerness to do something
2. Something that arouses a consuming interest.

The true fact of the matter is that the science community is aware of the facts in this case and in many other cases for the duration. The science community will not go out of its way in order to prove something that it does not agree with in other words the science community could also come up with the same facts that I have stated because they are very commonly known and I'm not secretive they are not hidden they are not placed aside under some box that no one can find most of the things that I have talked about and bridges on chapter 5 going to Chapter 10 these discoveries are very well known by a number of scientists halls of institutions of higher learning and great experiments that reveal the truth of my statements and many well learned in evolution science. X. evolution science and scientists around the world long before I put my thoughts to paper this is not something new or secret what is not done is this not made public by the science community most of the science community is well aware of the things that I talked yet and still they deny they lie and they stand by what they know. Just who is sick of the facts that someone else had said it first and that they would not stick their neck out to change what they know is true and they continue to come up with some idea or reason in order to stay or to say that they think they still have the answer of the origins of man which countless time after time after time as we talk about here in just my one small book alone tells continuously how all the pieces of the puzzle that they have put together if you examined under a microscope you will fined that they are not adding up to anything other than religious enthusiasm.

But we must conclude that science in the its infancy and the common man as non-compliance when it comes to the understanding of any parts of their work but if it walks like a dinosaur leaves footprints in stone when mixed with human footprints and these two events are proven by world paleontologists to be dinosaur footprint and stone guess what, then they are what they appear to be. This is not an illusion, this is not a deception this is the reality of what is being perceived, and clearly understood and this is not a religious experience because intelligence tells us that when the proof when the facts, when the evidence stands in front of us especially when we can touch it the PFE is true.

ICA Stones of Peru

A number of artifacts discovered have it made available over the past 20 years or so here is another. Some researchers as of late discovered a number of round stones and spheres. The fine was to be classified to date sometime in the dinosaur era as for when the stones and spheres were created. But that's not the interesting part the interesting part is that the spheres had drawings impressed into them pictures of surgical procedures and in detail, also flowers and a number of human interaction of things and events. The ones I found most intriguing were the ones with humans playing with dinosaurs this looks like a piece of artwork that someone had painted that was just another part of their collection or just a part of their life or just something that they have around the house as if it was another pencil or printer or light over dictionary or a calendar or anything very, but nothing unusual for the house and it was just something else around the house.

Logical commonsense question:

If humans and dinosaurs didn't exist at the same time, where did the humans that receive these sculptures get the ideas to add the dinosaurs?

If humans were not present or if dinosaurs were not present what do you have?

Lastly how did these persons (Humans) who had never seen a dinosaur draw one in detail the sculptor was playing with something that he had never seen?

This is not possible.

In other words this is not little Johnny drawing a picture of him self and his dog because he likes to draw with pen and paper or stone and rocks this is little Johnny drawing a picture of something that he is used to interacting with.

The question is this, how does little Johnny draw in detailed a picture about a dog if he has never seen one? Why would he draw a picture of something that he has never seen? Since these are adults drawing the pictures in detail as we understand they have more knowledge that little Johnny, surgical procedures in detail, different dinosaur shapes in detail, this was not little Johnny this was little Johnny's father and Johnny's father's friends in the medical bay or hospital also in the science laboratories or maybe even in the archaeology department or either in the school or the place where everyone studies? Either way someone was thinking and understood that sometimes you have to write things down in order to remember them as institutions and organizations do normally in any ordered society. Also because maybe paper wasn't invented yet and the only tools people had were walls of stone and stones of stone? I find it inordinate that sometimes only a few people like myself are the only ones asking these types of questions they are not necessarily difficult ones to ask you are not necessarily a un-capable asking us questions to answer they are in fact the right questions to ask and I do believe that they go on unanswered students who ask do not go forward with in their disciplines at the science secular colleges and universities.

Here's more the one verification over something different above that were is accepted by the science community another verification of man interacting with dinosaurs.

Remember I said that it takes more than one experience with experience in order to verify an experience this is the way we can verify, clarify situations and to view the only true things that we can hold onto and see with clarity this is right and true method and real.

So here you have another situation or proof, fact, or evidence, in order to verify or to disprove one ideology and to prove another's beliefs.

In other words have man interacting with dinosaurs which disproves the ideology of evolution and you have man interacting with dinosaurs was proof that man did walk with dinosaurs this proves that man a dinosaurs were created and design and living side-by-side as they were created by God a specific design to do a specific things. There is no such thing as luck there is causation to all things we just don't know what the causation always is but if we research you take time to look if we take time to stay most of the time we can find the causation to the center. There is no such thing as luck, and chance, a random selection, if that were the case there would be nothing but total chaos within our vernacular.

Logical common sense

This information is on display and is well documented and well known by the science community.

Logical, common sense:

If you have such a strong belief in something that you're willing to lie or denied knowing that the truth is out there. When you think so strongly about something you believe so strongly I think this could be classified under religious enthusiasm!

Enthusiasm noun
1 Passionate interest in or eagerness to do something
2 Something that arouses a consuming interest.

The true fact of the matter is that the science community is aware of the facts in this case and in many other cases for the duration. The science community will not go out of its way in order to prove something that it does not agree with in other words the science community could also come up with the same facts that I have stated because they are very commonly known and I'm not secretive they are not hidden they are not placed aside under some box that no one can find most of the things that I have talked about and bridges on chapter 5 going to Chapter 10 these discoveries are very well known by a number of scientists halls and institutions of higher learning and great experiments that reveal the truth of my statements and many well learned in evolution science. X. evolution science and scientists around the world long before I put my thoughts to paper has known about these things. This is not something new or secret what is not done is this is not made public by the science community most of the science community is well aware of the things that I talked yet and still they deny they lie and they stand by what they know. Just who is sick of the fact that someone else had said it first and that they would not stick their neck out to change what they know is true and they continue to come up with some idea or reason in order to stay or to say that they think they still have the answer of the origins of man which countless time after time after time as we talk about here in just my one small book alone tells continuously how all the pieces of the puzzle that they have put together if you examined under a microscope you will fined that they are not adding up to anything other than religious enthusiasm.

But we must conclude that science is in its infancy and the common man as non-compliance when it comes to the understanding of any parts of their work but if it walks like a dinosaur leaves footprints in stone when mixed with human footprints and these two events are proven by world paleontologists to be dinosaur footprint in stone, guess what, then they are what they appear to be. This is not an illusion, this is not a deception, this is the reality of what is being perceived, and clearly understood and this is not a religious experience because intelligence tells us that when the proof, when the facts, when the evidence, stands in front of us especially when we can touch it the PFE is true.

ICA Stones of Peru

A number of artifacts discovered have it made available over the past 20 years or so here is another. Some researchers as of late discovered a number of round stones and spheres. The fine was to be classified to date sometime in the dinosaur era as for when the stones and spheres were created. But that's not the interesting part the interesting part is that the spheres had drawings impressed into them pictures of surgical procedures and in detail, also flowers and a number of human interaction of things and events. The ones I found most intriguing were the ones with humans playing with dinosaurs this looks like a piece of artwork that someone had painted that was just another part of their collection or just a part of their life or just something that they have around the house as if it was another pencil or printer or light over dictionary or a calendar or anything very, but nothing unusual for the house and it was just something else around the house.

Logical commonsense question:

If humans and dinosaurs didn't exist at the same time, where did the humans that receive these sculptures get the ideas to add the dinosaurs?

If humans were not present or if dinosaurs were not present what do you have?

Lastly how did these persons (Humans) who had never seen a dinosaur draw one in detail the sculptor was playing with something that he had never seen?

This is not possible.

In other words this is not little Johnny drawing a picture of himself and his dog because he likes to draw with pen and paper or stone and rocks this is little Johnny drawing a picture of something that he is used to interacting with.

The question is this, how does little Johnny draw in detailed a picture about a dog if he has never seen one? Why would he draw a picture of something that he has never seen? Since these are adults drawing the pictures in detail as we understand they have more knowledge that little Johnny, surgical procedures in detail, different dinosaur shapes in detail, this was not little Johnny this was little Johnny's father and Johnny's father's friends in the medical bay or hospital also in the science laboratories or maybe even in the archaeology department or either in the school or the place where everyone studies? Either way someone was thinking and understood that sometimes you have to write things down in order to remember them as institutions and organizations do normally in any ordered society. Also because maybe paper wasn't invented yet and the only tools people had were walls of stone and stones of stone? I find it inordinate that sometimes only a few people like myself are the only ones asking these types of questions they are not necessarily difficult ones to ask you are not necessarily a un-capable asking us questions to answer they are in fact the right questions to ask and I do believe that they go on unanswered students who ask do not go forward with in their disciplines at the science secular colleges and universities.

Here's more the one verification over something different above that were and is accepted by the science community another verification of man interacting with dinosaurs.

Remember I said that it takes more than one experience within an experience in order to verify an experience this is the way we can verify, clarify situations and to view the only true things that we can hold onto and see with clarity this is the right and true method and real.

So here you have another situation or proof, fact, or evidence, in order to verify or to disprove one ideology and to prove another's beliefs.

In other words here we have man interacting with dinosaurs which disproves the ideology of evolution and you have man interacting with dinosaurs as proof that man did walk with dinosaurs this proves that man a dinosaurs were created and design and living side-by-side as they were created by God a specific design to do a specific things. There is no such thing as luck there is causation to all things we just don't know what the causation always is but if we research you take time to look if we take time to stay most of the time we can find the causation to the center. There is no such thing as luck, and chance, a random selection, if that were the case there would be nothing but total chaos within our vernacular. Good and riches comes to any one power, it is what you do with that opportunity.

Logical common sense:

This information is on display and is well documented and well known by the science community.

The Acambaro Mexico figurines

South America's also become a bit of great information and resources on the archaeological discovery venous

In South America some researchers discovered a store room apparently locked up by someone who was a collector of sorts.

There were a large number of boxes founded in the storeroom and they appeared to have been there for some time after the boxes were opened inside the boxes were filed with 30,000 different types of figurines of man interacting with dinosaurs. Dinosaurs of all types humans and dinosaurs of playful things in like manner similar to that of a man and a woman of today would pose with a bird or a dolphin or with a lion, tigers or bares, oh my!

These figurines had been tested and dated the interesting thing is to say they were dated to a time of recent time recent day and interestingly enough if the dates are you say 10 15,000 years old this may be close to the evolution idea however, this is still very young. People didn't make figurines and if they did they probably would not have lasted that long 10 to 50,000 years old. And they were close to 100 years old they would have been normal people making figurines? People collect figurines it is not unusual at all so if these figurines number up to 30,000 of them say were tested, categorized, separated, and set off to the lab to be dated properly the time of the creation of these figurines dated for 1500 to 4500 years ago. .

Now this is very interesting part, there were no major diggings, excavations, or fines or even pictures, of dinosaurs discovered by a man before more than two hundred years ago. No pictures before 20 century, no dinosaurs, no anything from which man's idea on how these fossils looked, there was nothing for man to reference from, to or anything. There was just nothing available. Given that the figurines were dated between 1500 years to 4500 years ago that is still within the top line of the creation of man and the relation of man and dinosaurs so here is another third reference of man interacting with dinosaurs. People have ideas but no one had ever seen any of the great animal bones that are on display presently around the world in museums today so what does this mean? This means that whatever date that was placed on all of the figurines was correct and man has multiplied by too many zeros in most of the past dating systems.

See if you want to believe something you will find ways to believe it and make sure that no one else can find your secret even given facts in your face even given proof in-your-face even given evidences that disprove your way of thinking your institution's way of teaching your colleague's way. You think that may turn the whole world around.

One of the problems I have is that most evolutionists are atheists and I don't understand why? Even Darwin believed in God there was a point when he believe that God created all things and then they evolved. At this appoint in his life this was changed later and then restarted later but I don't understand why a few believe that if there were dinosaurs then there could not have been a God big enough to create a dinosaur yet the same God created all of the universe all of the stars all of us all with everything that exists but not big enough to create dinosaurs? And again we have man interacting with dinosaurs how can men create a figurine of a dinosaur of all shapes and sizes in detail if he's never seen one? So man and dinosaurs being coexist together on this planetoid the same as all the other animals. A number of drawings of the stones of man interacting with dinosaurs and that's not the interesting part of the information in the picture that seems more and a attention getter is in the actual picture itself depicted its pictures showing the men attacking the dinosaurs but even that's not the most interesting part of this phenomena the most interesting scenery of the drawings is man attacking the dinosaurs in its most vulnerable spots in the same way a man would attack and kill a buffalo, a dear, or a bear, or any type of animal being hunted by a group of men on the Safari or the hunt up close and personal. What does this tell us? Simple that men not only interacted with dinosaurs man hunted them down for game, man ate them, the new men not only interacted with dinosaurs but also knew how to kill them and they did just that with in their neighborhoods. So this was not usual it was not dinosaurs 10 and man zero it seemed more of man 10 dinosaurs zero. It appears that man interact with animals in the same way we interact with them as to some tribes in Africa at the alliance of man only having spears in their hands a right of passage perhaps. A man who live in some tribes in South America will kill a tigers in the same way they used to because men even in North America was known for killing giant Kodiak bears which I am sure has the power to kill or destroy a great number of men at the same time. Yet and still man somehow, someway always seem to devour these large beast of burden to our way of life and these present-day animals are not much more antagonistic than the dinosaurs of past times and yet we have figurines and sketches of man interacting with them destroying them and in past times playing in having them for dinner. This is not usual this is not unlike man this is man saying thank you for the killer whale, thank you for the sperm whale, thank you for the blue whale with all thanks you forget the largest mammals on the earth the Tiger, the whale shark, the sperm whales and any other animal or mammal that has lived on this planet somehow, someway man has learned to conquer these and this is not usual this is normal. So how could science ignore several happening so easy? If you try it easy, if you have to lie it's easy, if you don't want it to be it's easy, if you want to see, it's easy to say there is no God, it's is easy just to be a fraud. These figurines are similar to the ICA Stones of Peru yet found in a different part of the world by different group of people in more than one experience with in an experience, someone sharing the same experience in a different experience within the experience this is verification in the third and fourth time of mans interacting with dinosaur. What does this, this means? That the evolution idea of the separation of man and dinosaur is completely flawed?.

Now someone wonder what happened to the dinosaurs? Evolution scientists speculate that a large meteor came from Outer space struck the earth and killed off the dinosaurs, some say that there was a great earthquake, some say that they just died or froze to death but no one speculates that man had some of them for dinner and that seems the most posthumous answer of all and no one ever stopped to ask the question what did the man eat? What was his digestive diet? How do you survive? What do you drink? What do you need to for support? There was no football, baseball, basketball; there was no Soccer, ESPN, no cable, no World Series, but there was hunting and catching, carving up and dressing for lack of a better term dinosaurs. Blue fin Tuna use to weight 1500 pounds a piece, per I've seen pictures of them huge monstrosities, just a generation ago, today there are only a few hundred pounds at best because of over fishing mans got to eat. He also did the same as he walked with the maga-li-saur-ist and you tell me that man didn't hunt dinosaurs! For me this is speculation but I think it's has some truth because if there were dinosaurs and man together walking, you better believe that man made a sport of this and them and hunting them, had them over for dinner!

So, logical commonsense question:

If man had never seen a dinosaur, never interacted with a maga-li-saur-ist of any type why would a man draw a picture in detail explaining how to conquer the giant beast? I would say to pass this on to other men and onto the children so they too would know how to survive in this life, isn't this what we do teach our sons?

Man of today always tend to think that we were the only thinkers and we are the smartest man to have ever walked on this planet above all those that came before us when I say that we only have more tools than they did, we are only more advanced because I do believe that given the chance to men of the past if he had our advanced tools he would do some of the same things and maybe even a little bit more given the grace of God hallelujah!

The metallic Grooved spheres of South Africa and the Bola Stones of South Africa.

I suspect that these were mostly used to attack large monstrosities in Africa

Metallic grooved spheres from South Africa are on display in a museum There. The spheres themselves look like a small round ball however metallic in design it has three protruded lines around its center that conferences it. Hundreds of these grooved spheres were unearthed by minors in South Africa purely by accident and they were perfectly and completely without a doubt that they were man-made. The spheres look like they were milled, smelted, or molded why I say they were man-made that is what I was told they are on display in a museum in South Africa you can look them up on the Internet. There are a **lot** of ideas about them however, there's one thing that's most interesting about them; they predate man, predate dinosaur, predate land, predate water, predate space, they predate anything in the universe that we know! These are facts present in evidence these on display in a museum in South Africa I am not the first to discover them. They have been well documented by science for decades. Understand if anything contrary to the evolution ideology anything different from their way of thinking if something doesn't match their way of evolutionary ideologies then they just discard them it doesn't matter the truth of the matter is it only matters what they want. This is my baseball bat and even though I'm the weakest player in the field and I can't play as good as anyone and I'm the smallest player on the field, I won't wait my turn I don't want to wait for the next game if I can't play right now I'm going to take my baseball bat and my ball and baseball glove and I'm going to go home! So says the -self-seeking, self-centered, only self concerned 12year-old child.

I don't mean to indicate that grown men and women of science are the same as any 12 year-old child but in some reasoning's it does appear to be so.

The spheres have been on display for almost as long as I am old and evolutionary science has always been made aware of them.

This is not proper grammar however, this the best way I could express what I wanted to say here.

The Bola stones of South America

The Bola stones are a perfectly round stones found in South America use by the Aztec Indians or some other primitive people they cannot be dated but they are older and date to be older than the dinosaurs they are on the Internet.

There is a large number of Internet sites and web pages dedicated to the art, archaeological dig sites on the Pulexy River Gulf at Glen Rose Texas, the ICA stones of Peru and the over 30,000 figurines found in the storeroom in Acambaro Mexico and the dinosaur art.

Evolution is an archaeological Pale ontological possibility

Chapter 6 The geological oil pressure reports

When you see something wrong and knowingly do nothing about it when it is with in one's power to correct the problem then we allow sin to prove it's self.

"Pallets"

On this planet that are laws that man must be abide by we are not always willing to obey these laws but we are aware of them such as driving on the right side of the road in some countries as are what we do in the United States we drive on the left side of the road in the United Kingdom these are laws that we abide by and the majority of the people obey and abide by them completely, these and many other laws and rules most people abide by. We do this without incident because it is our natural instinct to obey laws and rules set before us. There are of course the occasional slipups and slide downs in the course of the day but for the most part of the overall whole of the spectrum most of us are in line with the compromises that we will allow within our lives.

This planet also has laws that it applies by compromises that it adheres to, laws like the earth beneath us is constantly moving pushing itself against itself these are called platelets, this is happening all over the world this is one of the reasons that we have fault lines that are the epicenter of earthquakes. What happens at the centers this is the place where the earth has pushed itself against itself and the strain of it is so great that the earth will eventually give way what happens at these times is the pressure of the earth is released in the collapse of the earth after being pressured and pushed against itself. All of the eventually present earth disintegrates toward lower parts and everything within that region also disintegrates and changes its chemical makeup where it causes the same space to collapse and this could go on for miles and fault lines cover the complete earth above and water beneath. water at the bottom of the sea floor and on top of land. This causes the rest of the planet that's within the given area to shake and quake. We live on a very sensitive and unstable planet it's very unpredictable the slightest earthquake can destroy whole cities, can collapse whole landscapes this is where we live and we are used to living with it truly because we don't have any other choices. The ripple effect could go on for days and weeks because of aftershocks mini-earthquakes this is just a part of the overall whole. When we look at a volcano we do not see it moving from day to day until one day we see the top of the mountain law blow off however, we do not see that the earth was moving because they move ever so slowly. As we take pictures and as we look at the mountain if we could have a camera or cameras on the mountain we can see the mountains take different shapes from time to time over time and see when it's about lose its shape and never remain the same shape again. Because it changes shape over time ever growing outward at work preparing for the eventual and inevitable evacuation we really don't know when, how, or why, we only know that there will be an expulsion. As I'm writing this as we walk, as we talk, as we eat, as we sleep, the earth is always moving underneath us. As the earth moves the different types of dirt rock water and other materials in its vernacular earth also moves oil up to us so we can drive our cars. Of course most of the oil has to be pumped from beneath us up to us and we do this with drilling wells around the world drilling platforms this is what we do and everyone is aware of the business of oil drilling the from beneath the earth it's like bursting a balloon filled with oil or different types of crude as it is called sweet crude is what we use for our vehicles for the most part there are different types and different names we won't get into that what we will get into is that when we tap into these flows of liquids to be pumped up to us for our manufacture to be used as we will they have a shelf life and what I mean, a shelf life in other words there is a limitation on the amount of oil debts down there in certain areas around the complete planet so after a certain time we run dry in certain sites so we change and move our platforms to other areas to continue our excavation of fossil fuels from that level that us in order to fuel our lives. So now as I said from time to time the pressure of the oil dissipates and even the oil itself dissipates because there's only so much oil. The first area pressure of the oil has dissipated or had been pumped by a man-made source of to us.

What does this have to do with evolution?

The earth is only so big and only has so much oil and it only has so much pressure in the ground for which the oil will pump itself up without man's help if the planet that we live on head been here for un-toll millions and billions of years as the evolutionists claim the oil pressure from the oil and other minerals and a number of other natural spring waters, natural minerals that are beneath the earth that we also escalate, the pressure from these areas underneath us from oil and other materials would have dissipated long ago because there is only so much pressure on the earth, and any geologists can measure it and tell you they can also estimate the that this earth could not have been here for the number of years that the evolutionists want you to believe because there would be no pressure from the oil and other minerals to extract still from underneath the earth don't take my word for it ask your local geologist of which I am not I'm only stating the proof, the facts, and evidence.

The geologist the people that are trained to measure the Earth and all of its movements they go to school to develop a technique also in the area of date the age of the earth they learn the way to study how to tell how old the earth is also as a part of their discipline.

The geologist are educated to measure the amount of pressure that the earth is using to move the different materials and minerals and not only the mount of pressure but also the date of the pressure on how long the pressure at any given time frame has or was or will be used up and that there will be no more oil or pressure in order to pump the oil or minerals of natural gas to the surface this goes to the geologist being able to say because the earth has been here this time this is why the pressure to pump oil is very strong or very weak this is a normal part of their training this is what they do ask any geologists and from the ones that I've heard the words of training and truth came to me from them that if the earth was billions and billions and tens of billions of years old the oil pressure in order to pump the oil from the depths of our earth would have depleted and dissipated billions and billions of tens of years ago. So the earth is not as old as some would have us to think just ask any good well educated and experienced geologist.

The old geologists report

The new geologists report has always stated that measuring from the pressure and the great debt to all that has been accelerating from the center of the earth to the surface of the earth has never been dated any older than 10,000 years that's right never more than 10,000 years old so wait a minute if the pressure from the oil has been dated by all geologist to never have been more than 10,000 years old, how can the evolutionists say that this planet is billions of years old? Is someone making up the evolutionists reports does someone just believe that it is as they say, or is it a fraud?

Maybe it's a religious experience people want to believe something without any PFE. To believe something is a cousin to faith, faith in something without tangible proof, facts and evidence, is someone having an experience of believing without seeing a religious experience! Religion is okay one can believe in what ever one would like to one can be religious about anything that one would like to this is the right of every human being to choose what they will and will not do in any given situation we all however, must also accept the consequences of our actions. But we know that we can choose to do what ever we want to, believe want each will want. The consequences may not be what we want we can choose to do what we want so the old saying goes, 'Don't do the crime if you don't want to do the time' that's a valid saying in this case the evolutionists believe religiously that the planet is more than 10,000 years old with no proof, no facts, and no evidence to support their ideology because all evidence to the contrary supports the contrary. What remains by default is enthusiasm, religious enthusiasm.

The whole of the earth would have erupted the total surface of the earth would be moved alone the according to the science of studying the Earth by the numbers.

The evolutionists are notorious for adding zeros to the calculations why?

Some of the tests used to measure time and date are performed in the following and like manner a number is fed into a computer or a testing simulation or a testing site with a question, 'Is this piece of wood as old as the number that's chosen in that memo'. The number could be whatever number the evolutionists choose they always choose billions of years tens of billions of years in fact they could choose one or two years or 10 trillion years that doesn't matter they oftentimes choose a number that's in line with the ideologies as to support their ideologies to make their thoughts as they see it but in fact they are still not theories they're still only ideologies that just will not prove. The geologist's reports can't be as general because oil is a perishable and the mineral is a perishable that's used by the world the age is a factor in the process of the problem this also tells us how much oil is used in the search. How long it has been there and how long will it be there. These are no strange new way of thinking this is not something that is brand-new this is something that has been happening for lease one hundred and 50 years the geology reports have never been wrong once only the evolutionists guesstimations.

The laws of geology also have to be obeyed even when some of us don't agree with them.

The moon is a fine example there are lots of interesting things about the **moon** that the evolutionist do not like to talk about given the number of things that we mentioned above about gravity about atmosphere about life still a few things they don't like to talk about hydrogen oxygen, the global force of gravity to hydrogen, oxygen, no natural life giving force

Funny thing is here, with the two planetoids being in such a close proximity yet being so totally and completely different and unlike the evolutionist can't explain the how's and the why's of this anomaly, they didn't even try. I talked about the atmosphere before the moon doesn't have one in fact this evidence disproves the big bang theory in other words there should be life on the moon very similar to that of the Earth in every way but there isn't any life nothing but a bedrock, a dead rock without water. The lack of PFD indicates that there was design or a distinction made between the two bodies in space this also is a puzzle to the people of evolution.

It's solid!

This may not mean very much too very few people although, to allot of people however a Mr. Neil Armstrong said in 1969, "One small step for man, (while stepping down off of the latter of the lunar Lander) One giant leap for mankind. While the commentators were going wild Neil also said: "It's solid!"

Interesting though most people wouldn't think that this statement was one of a great moments in time just as much or even more significant as the astronaut's historic step on another space small body with in our solar system. We will have to look at the rest of the story in order to find out what the magnitude of these two little words meant in the world of science and evolution.

The rest of the story as Paul Harvey would so amply ex-spouse,

The Moon and Earth together go around the sun and 66,000 mph,

They (the moon and the earth) run into cosmic dust. The atmosphere of the earth burns up most of the cosmic dust it hits and the water and wind takes care of the rest.

However, the moon does not have an atmosphere, wind, or water. Therefore the moon collects the cosmic dust. NASA scientists calculated that the moon collected approximately 1 inch of deaths per 10,000 years of cosmic dust. And they believe that the moon is billions of years old they figured the moon would be covered in a layer of dust one-mile thick (which is why the NASA scientists designed the landing shuttle with large, wide pads and meter length sensors to dig into the dust that would tell the astronauts how deep the cosmic dust went).

However, when they landed on the moon, they found only 1/2 to 3/4 inches of dust on the surface.

This is more suggestive of the moon being only 6000 to 7000 years old. (This, coincidentally, is the estimated age of the earth biblically.) It's solid. When stated by Neil Armstrong, disproves one more of the evolutionist theories on the origins of man.

For some reason unknown to all this statement never made the evening news. The evolutionists were aware of these facts in 1969.

So, to continue to believe in such an un-proven, and a incomplete ideology such as evolution given the lack of (PFE) proof, facts, and evidence by default is nothing more than, religious enthusiasm.

zThe evolutionist are at liberty to believe in whatever they desire or want to, any theory, ideas, beliefs, imaginations, or whatever they shall come to with in their own vernacular.

However, without proof, without fact, without evidence, of any kind, the evolutionist should not be allowed to ex-spouse their religion to others with the insistence that this is the answer to the origins of man, sorry it is not, and not at the expense of the general public, the public school system and the rest of the intelligent world.

Evolution is nothing more than a religion.

Evolution is a geological impossibility

Here is a little something I looked up on the Internet just a little bit information just to show how easy it is to find what you want to if you really want to! I'm no genius if I can find it anyone can.

NASA and the 'Missing Day in Time

Did you know that the space program is busy proving that what has been called "myth" in the Bible is true? Mr Harold Hill, President of the Curtis Engine Company in Baltimore Maryland and a consultant in the space program, relates the following development.

I think one of the most amazing things that God has for us today happened recently to our astronauts and space scientists at Green Belt, Maryland. They were checking the position of the sun, moon, and planets out in space where they would be 100 years and 1000 years from now.

We have to know this so we won't send a satellite, up and have it bump into something later on its orbits. We have to lay out the orbits in terms of the life of the satellite, and where the planets will be so the whole thing will not bog down. They ran the computer measurement back and forth over the centuries and it came to a halt. The computer stopped and put up a red signal, which meant that there was something wrong either with the information fed into it or with the results as compared to the standards.

They called in the service department to check it out and they said "what's wrong ?" Well they found there is a day missing in space in elapsed time. They scratched their heads and tore their hair. There was no answer. Finally, a Christian man on the team said, "You know, one time I was in Sunday School and they talked about the sun standing still."

While they didn't believe him, they didn't have an answer either, so they said, "Show us". He got a Bible and went back to the book of Joshua where they found a pretty ridiculous statement for any one with "common sense."

There they found the Lord saying to Joshua ,"Fear them not, I have delivered them into thy hand; there shall not a man of them stand before thee." Joshua was concerned because he was surrounded by the enemy and if darkness fell they would overpower them.

So Joshua asked the Lord to make the sun stand still! That's right--"The sun stood still and the moon stayed---and hasted not to go down about a whole day!" The astronauts and scientists said, "There is the missing day!"

They checked the computers going back into the time it was written and found it was close but not close enough. The elapsed time that was missing back in Joshua's day was 23 hours and 20 minutes--not a whole day.

They read the Bible and there it was "about (approximately) a day" These little words in the Bible are important, but they were still in trouble because if you cannot account for 40 minutes you'll still be in trouble 1,000 years from now. Forty minutes had to be found because it can be multiplied many times over in orbits. As the Christian employee thought about it, he remembered somewhere in the Bible where it said the sun went BACKWARDS.

The scientists told him he was out of his mind, but they got out the Book and read these words in 2 Kings: Hezekiah, on his death-bed, was visited by the prophet Isaiah who told him that he was not going to die.

Hezekiah asked for a sign as proof. Isaiah said "Do you want the sun to go ahead 10 degrees?" Hezekiah said "It is nothing for the sun to go ahead 10 degrees, but let the shadow return backward 10 degrees.." Isaiah spoke to the Lord and the Lord brought the shadow ten degrees BACKWARD! Ten degrees is exactly 40 minutes! Twenty three hours and 20 minutes in Joshua, plus 40 minutes in Second Kings make the missing day in the universe!

References:

Joshua 10:8 and 12,13

2 Kings 20:9-11

Apes and humans?

Certain groups of people are always attempting to categorize all of the creatures on this planet in the same family, humans and animals for instance. We look similar to each other, we have some of the same characteristics however, this is where the similarities end and why we are always placed with in the same family to make things seen like one. Evolution is always doing this that is their way of trying to make their ideology, or their religion, or their recent reasoning more real than it is. But how is it that they always have to use different adjectives or different integers or something to fill in the blank of the space? This smacks and insults human intelligence in order to leave us with religious attributes. It is in these incomplete ideologies and theories that appear to be reality but in fact, they are only the perception of reality. They are only ideologies only what people think only what they have surmised by default that there is God. This would be a good thing if there was any validity to their ideology this would be a good thing if they had any real science to back up their ideas and their assumptions and there oscillations. However, evolution have had nothing they can look at but we can look at their ideas even more in depth under a microscope and when we enishiate a totally exam of things presented to the general public and to institutions of higher learning we say to ourselves at the final analysis we've come to an end of the story.

Belief, noun
1. **Acceptance by the mind that something is true or real often underpinned by an emotional or spiritual sense of certainty.**
2. **Confidence that somebody or something is good and will be effective.**
3. **a statement, principle, or doctrine that a person or group accepts as true.**

4. Religious faith.

The overflowing is the only finding that the people of evolution have come up with to date and the results are the same.

The evolutionist has come up with about nine different species of skeletons or bones or teeth or a member of the other artifacts in an attempt to justify their findings a walk of life. If this can be called,' A Walk of life'. This still has no validity or fact to be found within its walls. Here are some of the historical failed ideologies of the evolutionist that has been debunked over the years.

Lucy: nearly all experts agree that Lucy was a 3 foot chimpanzee.

Heidelberg man: built from a jawbone that was conceded by many to be quite human.

Nebraska man: scientifically built up from one tooth, later found to be that of an extinct pig.

Piltdown man: the jawbone turned out to belong to a modern ape.

Peking man: supposedly to be 500,000 years old but all evidence has disappeared.

Neanderthal man: at the international Congress of zoology 1958 Dr. AJE. Cave said this after his examination of the Neanderthal man showed that this famous skeleton found in France over 50 years ago is that of an old man who suffered from arthritis.

New Guinea Man: dates back to 1970 this species has been found and the regions of just north of Australia, who lives there the aborigines.

Cro-Magnon man: one of the earliest and best fossils is at least equal to in physique and brain capacity to modern man so what's the difference?

These are the historical foundations of evolution that they have been teaching in schools around the world for decades as proof that man has evolved through time these are the justifications that has been used by the evolutionist in order to fill the libraries of the world at the cost of untold millions of dollars in order just to say that they know how all life began and the origin of all things however, they are incorrect. Constantly the evolutionist has come up with a large number of puzzles and riddles and the dates that were as mysteries. However, they are not all accessible to the public. In order to keep them (the pubic) at bay from really asking the very serious questions from taking very deeply into the foundations of what the evolutionist are trying to portray in order to get a total results or a final results but the more you dig the more you look for one, the more you examine evolutions ideologies in its depth an objective mind, from an point of view, the more you find that evolutionist really have nothing of proof, nothing of fact, nothing of evidence, PFP. The evolutionist only have much to believe in much to suggest that could possibly happen, in what to have faith in, a religious experience about something that could mean a number of good or bad or inconceivable things that lead to a large number of things and assume that they or these, if they could or could not have happened over a course of time. However, since it cannot be measured by a man because they cannot probably and have not over the past 150 years. So many adjectives when they could just say yes or no, up or down, in or out, plus or minus. We are left with an idea that we have been sold a bill of goods and yet the bag is empty, other than a few adjectives and a few relics that they (evolutionist) believe in, has faith and tries to rely on that could possibly be some proof and is a test of some nonexistent epic because within the facts that are exposed, there is no depth no height of facts to the evolutionist, no summation. All we're left with is a religious experience a religion is okay one can be religious one can believe whatever one wants to that's okay there's only one thing,

Don't call your religion science.

I don't call my religion science I call my faith my belief in God that created me I say that science proves creation through conscious design, this proves that there was a plan, as I started to write on the page and not just to make lines of incoherent wavering that have no meaning at all I intended to write a book I concluded writing a book that was my direct intent that was the monetary result of the causation the result was my direct design and intent that I achieved by conscious direct enishiation and a specific goal I decided and completed from the first movement of my pencil being grasped by my hand.

Apes and humans don't mix

Isn't this evolution in reverse?

According to the evolutionists humans and apes are close to each other even cousins when it comes to relating to any other animal on this planet. When relating to other animals or creatures on the planet. The chimpanzee to be exact are our next primate however there are a number of steps between humans and chimpanzees if you look at the model of six the evolution's ideology there are thousands of steps between the chimpanzee and the man hints the reason the evolutionists believing that it took billions of years for beings to develop, however, the facts don't add up. As we covered in Chapter 6 the planet has not been here for billions and billions of years so all of this is fill in the blanks of thousands of missing links that evolution of this says did happen, however, this couldn't have happened here because the planet has not been here, all evidence to the contrary. Geology shows that this planet is no older than at maximal 10 to 15,000 years old at maximal.

So there would not have been enough time for all of the evolution to develop, at least not on this planet because the simple facts are this planet wasn't here long enough for those things to have had happened.

Logical commonsense fact:

most animals and all creatures on this planet are born by instinct walking, talking or eating, building, and running up to their mother for milk and they start from day one with in a few minutes or two and within hours, within a few days and weeks it's remarkable they start behaving like a bird, a lion, a snake, a fish, and on and on. Only the human, only would just lie there. The human being does not know where to go to get the fluids they need in order to survive. How to use the toilet, to clean him or herself, where to sleep to find his or her mother, who just gave birth to him or her, know to talk, walk or to do anything.

You will only lie there displaced and rejected because that is all you how to do.

The human being will only live there cry and with in 7 to 14 days die of starvation if left completely alone or if no one helped the child.

Logical commonsense question:

Isn't this evolution in reverse?

Humans don't mix with apes.

In the area of reproduction for starters, humans and apes do not reproduce in the same way. But shouldn't we if we came from them and they are part of us shouldn't there be some cohesiveness as the evolutionists say they are our closest cousins? We can mate make with our cousins! It's not right? however, we can't do this but, why not with apes? The though of this shouldn't be repulsive since they are our closes relatives right? I can see you all flinching in your chairs at the though of it! Why? Because our own very nature tells us that it isn't right! Because we are too different species created by the Lord God! Here's the thing, if we came from them then shouldn't there be some residual adherence to this and our linear reproduction in direct relationship to the chimpanzee to our cousin? Facts out there are? None. The human psyche is totally different from that of the apes we have features that are similar such as mouth, nose, ears, however, the shape of the heads are altogether different. This is no evolution at all.

The arms, legs, hands, and feet are similar but that's where the similarities end.

The animal world within itself are a different breed of animal. All animals on the overall whole are stronger then humans and a chimpanzee that weighs barely a hundred pounds can handle an adult human male that ways 200 pounds with no problem shouldn't this be reversed? Since us humans came from the apes a lower case of primate then shouldn't we be stronger and not inferior to the ape world strength wise? Sense we are the on and higher order of primates according to the evolutionary chart right? No other species on this planet can be considered equal to humans given all of the new advancements and developments humans have created on this planet. Anything could be considered, no species that has advanced the world as humans have, however no animals on their own volution have made an effort to communicate with us humans without the insurgents of the human beings instigation. Still there has been no progress. The animal worlds isn't helping or are unwilling to make peace with us it's still a war between them and us but back to the champs they are one of non-meat eaters. Normally they only eat within their diet of fruits and nuts, berries, when there are no other choices before them they still wait because meat is not their choice of food in fact if we normally see any eating meat of sorts we calculate that this particular animal is not normal and that and there is a problem with this apes digestive track or the apes natural instincts is displaying an usual characteristic.

Very few chimps get overweight, have heart problems, die from smoke inhalation, and get broken bones. Champs can on numerous occasions eat pretty much whatever they would like to and not get sick especially when they are found with in their natural habitats.

Resistance

Remarkably, the chimpanzees are very resilient when it comes to diseases; medical professionals have concluded research on the starting point of HIV and how it came to spread to people across the continent Africa so fast. This came about from eating ape meat. Digesting apes as meat is a very old and normal part of the diet for a lot of Africans. Yes some people to eat monkey brains even though, the apes can retain an enormous amount of diseases and proceed on through their life as if nothing is wrong like most animals lions have been known to survive a snake bite from the King cobra in the wild and live. The lion may be stunned for a few days or maybe even a week but these are natural events for the animal kingdom whose natural immune systems and different anatomy can resist disease and are far superior to the human being. and they can eventually become immune to certain diseases that would have normally kill them and in all cases eradicate the human being if they were exposed to the similar situations. **Evolution in reverse?** I remember seeing on National Geographic of a female lion receiving a snake bite from a killer snake the venom was so strong it stunning the lion for a number of days and the lion staggered her walk slowly for a number of days and was not able to remain within its normal habitat but continued walking and trying to survive. After a number of days or a week or so the lion survived and went on its way I imagine this doesn't happen to all lions but a human being would probably die within a matter of minutes from the type of snake and the amount of venom effectiveness of snake. This is the case in point is this humans is not as strong as the animal world? Not even our strongest people men, not even football players or the super athletes would they not survive a snake bite in the jungles of Africa, South America, or even Australia given the type of snakes and the venom that the most poisonous snakes produce that paralyze and kill humans within minutes? Well known document cases of course, documented cases of human deaths. The chimpanzees and other animals can live with the infectious diseases and different worldly poisons. The animal world does not blink at the same statistics of man when it comes poisons, diseases, infestations, and viruses, they normally keep right on thinking. They are like the Timex when it comes to these viruses and diseases they can, "Take a licking and keep on ticking!" John Cameron Swayze

So isn't this evolution in reverse?

You have to unlearn what you have learned in order to get past the brain-damage of what you have been taught.

Humans are a different animal

Human beings are similar to this kingdom the animal kingdom but we are not like them.

Take the chimpanzee for instance they are supposed to be our closest cousins' right? But what the evolutionists don't tell you are that Orangutan are more like humans than the chimpanzee in fact the word Orangutan means, 'The people of the jungle'. Things just don't add up. After some tests were performed on some other animals for heart transplants to humans only as an alternative as a backup for donor hearts for humans waiting to receive organs you know we have a real problem, because there are a lot of people that have heart failure and are waiting to receive a heart. And I can understand the research for the hope behind the want to sustain life. This is the only factor here and very rational how ever, I give you one guess about the results the researchers found inside their endeavors to retain a substitute heart. This is my major point to this whole chapter here the evolution say that our closest cousins are the chimpanzees and that's where we came from however, when looking for a substitute heart in the animal world when the evolutionary scientists decide to look to the animals of course they went to the chimpanzee however, this heart from the next primate in the line to us was not so a sustainable and adaptive to the humans anatomy. Even though, there was one other species that heart was more sustainable to the human biology was more much longer adaptable and more adaptive it was in the news a few weeks ago that Mrs. Bush Senior had a heart valve replaced most people didn't question the type of animal's heart valve that was substituted. However, this was the best substitute for the human heart valve and this one was found and it is being used in the heart and chest of Mrs. Bush Senior of whom I am sure could afford the best doctors in the world and they decided to give Mrs. Bush Senior who is alive and doing very well for the rest of her life only God knows how long with her new heart valve from that of a pig. Yes, the pig heart is more adaptable to the human anatomy. It is rejected the lease of all substitutes other than a human heart. Rejected the lease of any other species on this planet and scientists have known this for a very long time this is not brand-new if I am the last one to know this it is not brand-new. I don't think I get things firsthand I think I'm one of the last ones to find this out because surely I am not a scientist.

Isn't this evolution out of order? I mean if it's good enough for the ex-presidents wife should it be good enough for pretty much any one because after all I'm sure that they had some of the best surgeons and doctors on this planet operating on the ex-first lady.

Mrs. Bush Senior

The pigs anatomies internal organs were accepted, they were rejected less than any other animal in their species. The Kingdom of human and animal are not a like at all. The idea that the evolutionists have about our anatomies and the animal kingdom with most animals have more control over their bowels, on the man perspective I know most humans have a problem with the this and this is the reason why we take so many different types of drugs in order to help us relieve ourselves close to on demand.

Normally the case is this your pet is at the door barking or licking or purring at your hand trying to get your attention in order to open the door. Sometimes you get up right away and sometimes you don't venture, most of us oblige our pets. This is after we get up, get dressed, put on shoes grabbed the leash then go outside with our pet and allow them to handle their business by the next tree, or poll of or bush, or whatever they smell, or whatever it is they use. If we humans had to stand by the door and the weight for our handlers to engage within the same spectrum in order to adhere to us there would be messy floors and clothes to say the least. The point being animals have more control over their anatomy than humans I'm sure there are lots of people out there that have a pretty good handle on their ability to expel excrement but trust me as age and time goes on the controls will lessen. And I assume also for the animal world however their world is in a different place within both of our lives in that the animal world are still more efficient than the human with the ability to control his physique, even though, time catches us all.

Logical commonsense question:

What does the panda bear, the Zebra, the Orca whales in the Moo Cow, have in common?

Answer: they all have black and white spots and or stripes on them yes I know what you are saying, so? With all of these spots and stripes on countless animals and mammals one thing rings true what? None-of the patterns are the same on every new animal not only on the zebra but as well as the Orca on the cow (dairy cows more likely) on the panda bear yet they all seem to look alike but each and every one is distinct specific intended. This applies to any creature that ever has been on the planet, no two have even been the same. How can they all be so different? They look like the other to be a twin but not to be a twin and only to be a friend? And how did they get the spots and the strikes they evolve from something to what? If they did indeed evolve, shouldn't they all be exactly the same? Like I said earlier every ear is different. The two ears on the same head are not the same and not like any other two ears on any one head that has ever lived, has ever lived, that ever will live or that is living today isn't this an amazing? And one more question why does these particular animals and mammals have stripes or spots? There is no reason for a tiger to have stripes a lion has some stripes and spots but why? These are exterior yet they go to the skin through the hair of the animal? Is there a reason? It looks like a specific design to me, The Artist panting.

The evolutionists have no answer.

Animals don't have spirits

Lastly it has been proven that animals don't have spirits they have life, but they don't have spirits experiments have been conducted in this area for a number of decades. Animals have souls but not spirits in a number of documented results that were made available to the public they are part of record. There is one study that I have personally seen in the movies back in the 70s I saw this film in title, 'Beyond and Back, a doctor conducted experiments on 15 people and 15 dogs and cats at the same time of their deaths to prove this as illustrated and documented always with the case of the humans at the time of death there was a weight loss of about 2 pounds or so. However in all of the cases with all types of animals including dogs and cats or any type of animal that was found at the time of deaths there was no weight loss. Some conclude that this was the spirit of human beings the two pounds or so that vacated their body at the time of their deaths. Does this tell us there's no dog and cat heaven? Maybe, I don't know? I only know that there is a distinct difference in weight loss at the time of death between human beings and animals this is another difference between the two of us animals and humans that shows that we could not have come from them nor them down from us. We are two distinct species created in time to coexist on the small plan.

There are thousands of cases of people that have died and returned to explain in detail of what happened to them after death I don't recall any of them that I have encountered in my studies talking about animals in the afterlife. These were also well documented events and situations so much so that Oprah Winfrey devoted a complete show to this topic in the 80s I believe this also can be verified a complete show about people that have died and then returned in order to talk about what they saw and came back to tell the tale these were considered as spirits or their spirit that died and came back. This is old news, what is my point here?

Logical commonsense question:

Not one person in these cases or situations where someone died and returned has talked about evolution, no one who has to died in return has said , this was the next step in evolution. Granted now everyone that died didn't believe in God? No one. Thousands and thousands of cases no one has ever said anything about evolving. Given the fact that some people don't believe in life after death, this is true with the evolutionist however, this is a natural part of our existence. Also for all those that may have come back from the death none of them ever made statements a testing to evolve in some other place they were also aware of their existence where ever they were. Lastly if we can die and come back isn't that against the laws of evolution? Shouldn't we after we go there stay where ever we go to the next level after this level and not be able to return to this level? In other words we are evolving forward how can we come back? My point exactly is this when you examine this a subject as deeply stirring as evolution it promotes its self as this is something that is a life development, a life seeking answer, a life asking for ever question, and we as human beings on this planet should be able to come up with an answer given that we find answers for all points everything and everywhere on our planet it doesn't seem such a difficult obstacle to me. Think one thing, completed it go to the next thing.

Humans from animals are very far-fetched:

Lastly, the idea that humans from apes is very far-fetched there are thousands of steps that would have to take place in order for a human being to have come from an animal. Take for instance the respiratory system changes, the heart lung capacity alterations, the brain function adjustments, bone density increase and decrease, or ration function and motor functional skills, and these are only a few of the thousands of adjustments and movable body functions that would have had to be considered in order for evolution to work there are hundreds and literally thousands of misplaced anomalies it was, just count between two or three of the species and none have been proven to date no not one. But like Darwin said,

"Noting the abundance of fossils, numerous transitional must be found to prove my theory."

The evolution has been coming up with two to 10 different things that they have been putting together to try and say that this is proof of the transitional. It would've taken literally tens of thousands, hundreds of thousands, of different transitional in order to complete evolution over billions of years tens of billions of years. The evolutionist only has a few hundred pieces of matter and we humans and they say that we evolved? The facts of none of this not are evident. Now they believe, they have faith in what ever is they believe in? They can have faith is a good thing so is religion it's always good to believe in something but without truth, but without proof, but without facts, but without evidence, you only have enthusiasm, religious enthusiasm.

The humans and the animals are not as compatible as evolution would have the rest of the world to believe this is their religion. This is what the evolutionists believe and their sad attempt at filling in the gap between humans and animals. The truth is if you don't have pasta you can't make spaghetti. The evolutionists insist at an attempt to intrigue us however, to take their idea's old and new on how life started on the planet base solely on faith alone?

Sounds like a religious experience to me? This is something that I may have some understanding.

So don't believe the hype about evolution the facts just don't add up.

Evolution is Homological and Anima-logical impossibility.

Yes

CHAPTER 8   FOR 150 YEARS

Man's New World

All of creation has evolved from the levels that were to our advanced stages.

This is a summation of what the evolutionists are claiming to be the results of our move so much for work with man's world, idea, travels and the like.

For more than 50 years there has been a gentleman speaking the good news to the world. The distinct gentleman man who would not say of himself that he should be so highly esteemed but when you have been minister to the past 10 presidents, count them ten, that says something about you. One can like or dislike this person but one would have to at least listen to someone who still maintains this kind of Accolade and adoration.

This now gray-haired gentleman doing what he does on a crusade exposed this statement some years ago as a truism of fact into the consciousness of the existence present, pass, and future.

"For less than 150 years man could only travel as fast as a horse would take us."

This is a very profound statement given that man has traveled past the speed of sound, past the 22,000 mph speed limit as the space shuttle flies and with ideas of traveling pass the speed of light within his mind's eye.

This would seem normal to the man of today but there was a time when the general public with their fearful thought and the prospects of the city governments installing electricity inside the building and definitely, definitely not in the home for fear of electrocution although there were some accidents at the first installments it took a little time to get this new idea to work.

So mankind has surpassed the idea that we are limited to our way of thinking our new attitude is that one of we can do anything if we put our minds to it. This way of thinking seems to be true. Why? With the surgeons of the computer man can almost achieve anything for the comforts of mankind. Men all over the world in the last few years are experiencing vast and enormous amounts of knowledge that is propelling man to places that the Romans, the English, Egyptians, Asians, the Aztecs, and the Africans, of the past could only dream of. In this mankind has increased in the way that we do almost anything but is this evolution or is it intervention from another source?

If all creation is in the process of evolving to this upper level where we do things a lot faster this must mean that we are moving forward to the next stage in our evolution this seems right, right?

Logical, common sense question:

Why haven't the animals and trees, the planets the worlds and everything else, for that matter, why haven't they evolved? Only the humans are thinking faster and traveling more than humans over the past 150 years?

No cat's mouse trap

The male lion still gets his dinner in the same way that he is used to the same as 2 to 3000 years ago even up to 6000 years ago or more. The female lions hunt the food down the males take the food from the females and eat until they have had enough and the females eat what is left, unless there are no males around then the females eat all. Sometimes the females run the males off but not most of the time. The system works for the pride of lions and has for as long as man has known about lions. The male lions in turn the projects the pride from all other predators the males has his part in breeding with female lions and also the males have a part in the upbringing of the lion cubs.(The human female is living a lot differently now a day). In most of our cultures and subcultures on this planet but times are changing!

Logical common sense point:

Isn't all of creation, all of life as we know it, all of the animals, all of the trees, all the birds, all of the bees, are they not going along with life as we know it as they have been going along with long as they do with for as long as man could understand and live a life? In other words there has been no change in the rest of the planet in the way everything else lives. Live or have babies or each or walk or talk or communicate no also even in the plant life it won't grow any faster necessarily on its own. In fact truth is we can't grow the same amount of food for the world that we did 50 years ago. The truth is only man has changed nothing else even man has not evolved only change. Where is the new cat's mouse trap? Where, is the Wilder beast escaped trap? And the like, there are tens of thousands of wildebeests on the African plains and the other large predators catch them, they appear to be also at the beck and call of the large predators as are most animals' life on the prairie (Serengeti) in Africa and other areas of the wild animal kingdoms. Even though, man has gone from riding a horse to riding in a car, from flying a plane from his hand, to flying in a plane, from rowing the boat, to drive a speed boat, from flying in the air, to fly in space. All of these things man has engaged and we have evolved within the last 150 years and according to the evolutionists all of us are advancing all of creation so why hasn't the lion built a better way to catch the wildebeests, (the wildebeests gets away most of the time) why haven't the wildebeests found a better way to escape? Why haven't the birds and the bees found different ways to evade each other or the honey bee found a better way to make honey even though, the honey made the way it's been made for all these years is still very good. Even though, some of the bees have disappeared? Global warming. Why have the fish in the sea not found a better way to evade Wales, lobster, shrimps. List goes on since all of creation are moving forward and are evolving why doesn't the grass grow by itself perfectly? Why must we use fertilizers with different grass seeds and different lawn services in order to have beautiful southern grasses? If they were evolving should they do these things on their own? How is it that only man seems to be moving forward alone? Well, the Bible says, at the End of days "I will pour out my Spirit upon all flesh and your young men will dream dreams and then the old men will see visions. Also knowledge will increase and travel will increase." Men and women are having dreams & visions of all types. However the animals are not saying anything and yet men and their dreams of turning them into reality. However and elephant or tree or anything else in nature of some sorts are all growing at the same pace unless man helps it. All is still the same. I had a tree die in my front yard for the life of me I can't figure out why? But since I'm talking on the computer and writing sentences and paragraphs to be included in a book as a way of developing myself or expressing myself I am moving faster. In fact thank the Lord God that I know how to use a computer given that when I was born was the first year computers was created and yet today the computer are more people friendly than it was all those years ago and we have made enormous progress in the last years alone. Much more than the prior 2000 years, before the last 150 years man has not just turned but has turned extremely right and up and increased in knowledge and a large number of ways and in almost everything that he has put his hand on to do. Why have not the animals, or birds, and bees, the flowers, and the trees, also progressed to a higher level of advancement may be because we didn't evolve and they cannot evolve. Because there is no evolution there is only a direct specific intent for our minds and our beloved, His Holy Will! there is no such thing as luck, there is no such thing as accidents there is a causation behind every situation it is to be revealed the answer of all the things presently we can not comprehend because there is something above our comprehension we only use 10% of our brains capacity this is well documented and well known that leaves 90% of unused space.

Logical common sense question:

Where are the animals and plants the waters and the winds of flyers and all of the rest of creation's movement forward? Where is it? Please? To believe that evolution has worked or is working, please can someone point me to it I do not see it so it's not happening? (All of creation is moving forward to the next level of creation) then where is the rest of the working to moving the evolving I have not seen it and I have been looking.

Chapter 8

The world is breaking up all around us as we speak.

The idea of evolution says that one can't see the change in the middle of the change?

In other words because evolution is so big and so vast and changed ever so slowly that because we're in the middle of this gigantic change all of these changes are constantly happening around us regardless of what we understand that we can't see the changes are still happening around the sky like growing you can't see a little boy grow from 5 to 10 years older or from 10 to 16 years old day by day. However you can put a camera on somebody and watched them grow you watch a flower grow if you want to. I can see the universe I can put a camera along the universe and see it changed see it evolving? What, no? I understand because if it was evolving we would see something in it beause we are not seeing anything. But this is one of those things that evolution always says you don't have to see it to believe it but it is happening all around you believe it or not. Yes this is one of those things, looks like, sounds like, he is like, it must be, a religious experience.

Religion is okay with me you can believe in what you want to believe in, one thing though,

Don't call your religion science

If you have been living on this planet for the past 50 years or so even the last 20 years you have to admit that there has been some changes, changes that affect the way we do everything. Changes all around us we've changed our way of thinking in a number of different ways humans on this planet as we have changed our ideas about large number of things in a large number of ways some of us are still in different archaic ideologies but the vast majority of the human beings on this planet have advanced forward in thinking and ideas and realities. The ozone layer that protects us from all the rays of the sun that would destroy the people living here on this planet is depleting and not repairing itself it's interesting,

Logical, common sense point:

Without the ozone layer and the specific layers to block the rays out and allow other specific rays in.

Logical, common sense fact:

Life as we know it could not and would not survive hours on our planet without the ozone.

Man has a great part in the deterioration of the ozone's depletion.

Logical, common sense question:

Shouldn't the planet make adjustments for the forward movement of man? Shouldn't the planet since it's evolving anticipate that man would develop these things that were destroy the ozone layer and compensate in a way so that we could still go on making and using and developing fossil fuels as it distraught different parts of the creation yet still we should be able to survive because evolution is not responsible for the destruction of the planet evolution is only be responsible for the forward progress of the planet so if we are destroying the planet evolution help the new species that would emerge and evolve and the new era of life would spring forward and the ozone layer would just keep regenerating itself to supply this one and only planet that we know of that has also layer. It would keep supplying the layer so that the inhabitants of the planet would automatically be safe for ever they would never have to worry about the rays from the sun that could harm and kill us within seconds. The evolutionist would never have to worry about those things because after all this is random selection only the strong survives, only survival of the fittest so these things would help us to survive the planet with in of itself create these things to help people to survive regardless of what's happening in outer space on this planet because we have evolved in the planet evolved taking us with it and we are evolving to a point of safety because we are presently safe are we or are we?

I'm sorry I don't think so? That the planet is breaking up all around us as we speak that is the one part of this evolutionist plan for the planet that they don't talk about. See in every part of this planet there is a earthquake, erosion, there are floods, and mudslides, volcano eruptions, and the like in our present history there is starvation there, is famine how was it that evolution has not helped us to become immune to the diseases that are attacking most of the people on this planet? You have to look at what the evolution saying, ask questions. Remember they say that we have come through hell fire damnation, through the deepest of oceans, we've come through fish, snails, mammals, animals, they also say that even the cows went back to the water. But yet and still diseases and viruses are killing us, earthquakes and mudslides will destroy us. Even though, didn't we evolve through worst things worst things already if we involve? And yet these things are try to eradicate us from our evolved home? Isn't there something wrong with this picture? Someone please answer me because I know that I am not the only one out here asking these questions. At our present history there are more violent occasions on this planet than in the past 100 years then there has been in the previous 2000 evermore than ever before but man is still getting more knowledgeable and traveling faster. Last but not least man always has from the very beginning and is continuing to fight his brother for his land, is his food, goods and whatever is in the way of his cause at the moment how is this evolving?

Shouldn't man be evolving past the point of devastating his brother for mere pittance yet that is the one thing that man has been consistent at from the very beginning of the very first deviation and disrespect of one's own family and friends turn them into his enemy? How is this evolved sensibility? How was this event advancement?

Man the only change.

Looking for the advancement of the movement of the change some proof, some facts, and some evidence, of any type? I haven't found any I hear a lot of noise, sounds of yes, it is a, could be, that maybe, be this, yes, we discovered a new fossil and yes this means that we did evolve from something to something less we forget about the disease, the famine the wars, the earthquakes, the mudslides forest fires, what about the thunderstorms that are continuously increasing year after year, the flooding of cities, a destination of towns is the overflowing of the river banks the parts of town that are now under water that never will emerge again to violent destructions of different parts of our world paragon and would never return where is this evolve sensibility in this? It seems that evolution is only good for those that are evolving to something but not enough for here but where is the evolving here since all are here and we all are part of this great evolution aren't we?

The only consistent change that man is making is that more of them building weapons for mass destruction in order to eradicate each other one way or another. For what, to threaten, to eradicate each other one or another and this is forward moving, thinking? Sensibility to my understanding evolution is a progressive better way of thinking that bring peace and tranquility to peoples yet and still most peoples are arming for war. Nothing but room for bloodshed, this is not a joke people's lives are being devastated this is not a laughing matter soldiers going to war little boys coming back from war now old men all realities, lives that are changed forever tens of thousands of people dying.

Where is the evolution and that?

So the only change has been man if you look at the true facts the only change has been man. Nothing thing else on this planet has moved forward or backwards other than man who can't really say that mankind has moved? The over all knowledge has increased and we travel faster to think otherwise is to engage.

Religious enthusiasm.

Evolution is an un-movable impossibility.

CHAPTER 9   THERE ARE NO MISSING LINKS TO BE FOUND

Faith, noun

1. **Belief in, devotion to, or trust in some body or something, especially without logical proof.**
2. **A system of religious belief, or the group of people who adhere to it.**
3. **Belief in and devotion to God.**
4. **A strongly held set of beliefs or principles.**
5. **Allegiance or loyalty to somebody or something**

Blurriness

When a subject of interest hits you a light will come on. A picture will come into focus in the mind. Now with most people if the picture is kind of blurry no one would have to work on the point to bring the blurriness to fruition. Depend on the person and the subject this could be an endeavor of a great magnitude that could take years to arrive at the subject that could lead to a lifetime of research or for some it could happen in an instant.

Some 35 years ago a light went off in my head about the idea of what's in this book I didn't know at the time that I would be so interested in everything that I've been writing about at the time I just know that a spark was started and I haven't been able to turn it off for all this time and it is still on now and my eyes are still open my mind is still open my heart is still open and I still like to research look up topics in order fine a clearer understanding because is remarkable what you might get do when you put your mind towards something of interest, something that one really likes. I would be this far involved in the subject 35 years ago but I still talk to friends I like to watch a lot of movements I still read some books on the subject the interest. I still keep coming up with more evidence of how to disprove the ideology of evolution that part always astonishes me because I keep thinking that maybe I'm wrong sometimes and then I turn around and I see how evolution has shot itself in the foot one more time.

So when I was told that evolution was this way or that way I would check the way out so that when I would talk about the subject I would be correct to my annunciation the subject of missing links would come up on occasion I for one couldn't find the missing parts of people in the area of study continue to say we will find these links this is what Darwin was saying remember unless I find the transitional I need to find these links. I don't put a person down for trying to research and discover something that they really have a fire in a building their mind this is normal as whatever one does I also agree with Mr. Darwin when he said he must find the next step and order to complete his experiment because he wasn't of finding any didn't find it at the end of his research at the end of his life he did not find. The people that came behind him is taking his research way past what he even imagine with out fulfilling the end of the line in other words they have not found the missing links also they have not found all these transitional also yet and still they claim to have what he had not said that he had that is this great vast push all of creation that moves itself forward to the next step of evolution that just does not happen that just is not happening! Like I said before when you begin using a microscope you total up with one of two things that will come true, 1, the true story and 2, the store that is told that is on everyone's mind. In other words you could look at something with an idea in mind of what you want it to be and if it's close to which you want it to be thin that's exactly what you want to be but, in fact the truth of the story is its nothing like what you think is? It only has a little bit of green in the corner of the picture but the overall picture is red so you can't say it's a green picture you'd have to say it is a orange picture with a little bit green in the corner. This is what failed evolution is, somehow all of their ideas and ideologies are all true yet they have no PFE of truth to have a few pieces of bones and skulls and a few fossils yet they don't know how this thing is supposed to work. You can look at a car from the outside and it could look very well but, with no working parts or motor, no drive train, no axles. The wheels look nice and shiny the car is a pretty glossy paint finish everything looks very well, he opens the car door to find there is no seats, no dashboard, no steering wheel, it's all empty an shell. You have to look behind the door, you have to open the book to see this not just look to cover sometimes there is nothing behind the book cover than just the cover. One other thing, a lot of people today are dealing with one or two areas also; reality and the perception of reality a lot of people don't want reality they want to perception of reality which is easier overall but this is nothing more than a childhood fantasy of what one wants and that is reality to be for that person. It's kind of like smoking marijuana; you escape the reality of present day into a subculture reality this does not equal to the reality of real time. Some live this life it is not a part of real-time and it is a total subculture of that it is far away from everyday life. We can feel it with life as if we are in a total different world but in the mind's eye you see nothing wrong in your living in your perceived reality. One have only transferred from reality but it is not a real tablet. Watching television and all the great graphics that television and computers can make, you see some great graphics, there are some good ones but those are created inside the computer inside of your television even inside of your radio those are fantasies and they have no validity out here in the real world they are not real they cannot happen here and they are not a part of what happens outside of the cyberspace, the TV space. It is remarkable the fantasy of the star movies the people acting out in the movies they have never been in spaceships, they have never been in space, they have never known what it felt like to fly at light speed, they have no concept of weightlessness, they have no understanding of what it means to travel from one planet to another from one universe to the other but yet and still, if you watched the television shows you would think that they have a great idea better or at least equal to the NASA astronauts. When in fact they have no concept other than that one of reading from a page that someone else typed for them to read? It is not reality; it is the perception of reality the same holds true with the evolutionist. When we place evolution on the reality table and under a microscope we see the similarities with reality however, more so than not evolution is seen as compromise to the perception of reality.

Let's take a look

If this is evidence

Okay so after looking at a number of areas we find no evidence of missing links because they're none to be found, there are no fish with half gills and have lungs, humans never ahead gills the three little bones in your ear remained the same size from birth to death. The mammals such as Wales, Dolphins, mate frontally but this is a natural for them and has nothing to do with missing links. None of the findings above are missing from anyplace because they are where they are supposed to be thats normal for the creatures that live on this planet that have died from the natural course of time for all of the different steps that are claimed to be a part of what evolution base all of its ideologies, this would take literally thousands of creatures in order to filled the void not having been produced that have not proven to be credible? None! This scenario about fish coming to land and developing lungs from gills this has never happen and without this happening the evolutionist have nothing because if the fish did not come on land then you don't have a man homeland. The evolutionists really have nothing because there is no proof, facts or evidence to verify that any fish has ever Metamorphose a size into a reptile at lease not the type of Metamorphose that the evolutionist are talking about. Not at this present day in operation. There is no working model; there are no examples, of a fish developing lungs from gills in all of Earth's history this has never happened. So now that you've come to the end of your search and you have found the reality that you were after and now that you have found a different reality what do you do? Be politically correct? Or do the right thing?

Faith is the evidence of things not seen.

Evolution is an un-provable possibility

CHAPTER 10 NOTHING MORE THAN A RELIGION

Religion noun

1. People's beliefs and opinions concerning the existence, nature and worship of a deity or deities, a divine involvement in the universe of human life.
2. A particular institutionalize our personal system of beliefs and practices relating to the divine.
3. A set of strongly held beliefs, values, and attitudes that somebody lives by.
4. An object, practice, cause, or activity that somebody is completely devoted to or assessed by.
5. Life as a monk or a nun, especially in the Roman Catholic Church.

Choose a path in life

When we choose our path in life we can do pretty much do what you want to do pretty much what we would like to in life. the same way people choose to do the right or wrong thing or to love someone or to whatever for whatever reason not to love someone the same as with religion we can choose whatever we would like to believe. The dictionary gives a definition of words given with the language or language that's mostly shares a moment to list answers and explanations for the word religion also means something that infers similar to this paraphrase, 'Man's search for God'. In the end you know that's a good thing men can search for whatever he or she would like to attach itself to Evolutionary science and to claim to be a valid member of the community however, without PFE of so many missing links this strains human intelligence and the souls to common sense and logic of the mind in other words this appears, looks like, walks like comatose, and is asking all to believe and have faith in this way of thinking like a religion!

Now a day we know the people of the world believe in or have faith in what ever they want to. People have the right in order to learned that one can believe in something that are adamant or in an adamant that is something living are not living we are free to do what we will and more so when it doesn't hurt anyone and especially when it doesn't effect our closest friends, life, family, in any normal ways of walks of life. I don't think that most evolutionists have gone out of their way in order deceive the populace but I do think the group as a whole will not allow any new ideas to interject knowledge that might change things.

These are the six principal areas of evolution this is what they based their theories on and their complete ideologies.

1. Cosmic evolution, the Big Bang makes hydrogen.
2. Chemical evolution, higher elements evolve.
3. Evolution of stars, and planets from gases.
4. Organic evolution, life from rocks.
5. Micro evolution, changes between kinds of plants and animals.
6. Micro evolution, changes with 10 times.

The first five are not a science they are a faith this is part of evolution that all must believe in Evolution. One must have faith in this because these areas are not proven to put it more plainly no one living now in the 21st century lived thousands of years ago let alone in more tens of thousands of years before or billions and billions of years ago in other words the only truly and actually way to prove something in history is to have been there. A large part of evolution is based on assumption or faith. No one living here and now was there! So science must assume, assume to have faith in all of the people in this science. They are all taking by faith that all of the people, scientists, and instructors, before them. A large number of evolutionist is don't even know or where to find Charles Darwin's original notes, papers, findings, and study material (which is on public display) or his prime theory, or know what Darwin's ideas are conveying. So with the absence of proof, facts, and evidence, and no actual working model in order to prove any of the ideas above one has to take these principles by default on faith!

Most of the students have faith that their professors and instructors know what they are talking about. The theory of evolution is less than 200 years old. They all have to be assuming. Before Darwin a few people had different ideas on the origins of man some of them were atheists however, of the great minds none would even think to the level that we can come up from animals. Not the Egyptians who I might add were known to have believed in and worshiped pretty much any and everything that came across the table, nor the great philosophers, or architects educated in Egypt and returned to Greece in order to build the great Parthenon's and cities of Greece and Rome, such as Athens, not Alexander the great conqueror of the known world of his day, none of the great Caesar's of Rome that rule known world of their day, nor Michelangelo, Leonardo da

Vinci., Galileo nor Asians, the Aztecs, or the Africans or anyone on the planet ever thought in this way. Why? The people above back in their day the animals were inferior and to be risen up from animals was not something that the intelligent man would even allow to enter his mind. This belief of man coming from animals is 18th and 19th century thinking but still no one of us was there then that's here now. So whatever is presented has to be taken on faith!

The Number six principal theory of evolution

When compared to the genome this is not constant with the genome code we can start with a sexual procreation; first a sexual procreation and reproduction is not a strong enough linear system to support the upper levels of life as we know in the past the lizard stage, this has not happened with in our existence in theory yes but in actual living beings and practical working models there are none.

RNA ribonucleic acid

One more logical, since question:

What is Ribonucleic acid? Answer this is the idea of how all of this evolution got started in other words this particular type of acid is it every human being in all of everyone's makeup this is what the evolution say oozed out of the bottom of the ocean however, there is just one little noting in the way. None of the evolutionist in the whole world has a complete theory of RNA behind all of the steps, behind all the leaps, behind all of the migrations and the books, behind the libraries, behind the gills to turn to lungs unproven, ideologies, behind every single thing that the people of evolution have written and have filled the libraries of the world's behind all of that you will not find one book, one paper, in one library, or in one syllabus, or one piece of literature that says to complete theory of RNA. Why?

Because there isn't one! No **evolutionist** has ever come up with a working model. Not in all of their experimentations and why? In all, their speculation, and all of their adulation, and in everything that they say, No **evolutionist,** not one can prove how it works or if it has ever worked if it is working today, or if it ever will work. They have no way of proving their ideology. What does this mean? Out of all they have done and said through the years, **this means that they all have no idea how it all got started?** The evolutionist say that the RNA oozed from the bottom of the ocean somehow into the ocean and became a amoeba then, became a fish, that became a reptile on land no doubt, Olay, but they don't know how that happened they have no idea how that happened to happen and none of them has a complete theory of how that might have happened? And since no one was there billions of years ago, so it never really happened yet and still To me, they based the whole of their religion on nothing! Like I said they have an empty car shell looking good on the outside but on the inside there is no seats inside, the door, no floor, no metal axle behind the wheels, there is nothing but an empty shell and a they have no idea of how this all got started. How did I find this out I started asking questions and looking into what they actually believe and how did they come up with their ideology. Evolution is a limited to the perception of reality and not reality they sale you a chicken without a chick. The people of evolution have a noting of there own.

As I stated before the evolution says that everything started out this way; for no reason and out of nowhere all of a sudden, with no causation there was this big bang the power that could be put in a tumble exploded in our space(but there was no space) and pushed out all the planets of the stars and all the heavenly bodies into their complete orbits as we see them today all the planets were hot and mold lava over the course of time they cooled(who was there to see this) but after the cooling somehow, someway RNA all the makeup that's it every human being, every animal, every tree, life everything thing, oozed its way from the bottom of the ocean floor and became, an amoeba, and then, became a fish, and then it became a walking talking fish, and then it became a lizard, then it became a mammal, and then it became the animal and then became a man. And here we are walking around and talking because we evolved from nothing clearly by no choice clearly by chance and random selection we are here maybe because we willed it, we who work nothing will to be something? I don't remember doing that making that decision and I think I would've remember such an idea is that. And for the most part neither is anyone else on the planet living now pushing with the same force as the amoeba from the oceans floor to become anything this is all a fraud and someone's imagination sounds like Hollywood. I like movies I like the part where I can go out and come back in again it was fun I see lots of movies. One thing about most movies though they are the perception of reality, not reality.

It is remarkable to me how a group of intelligent scientists have come to the table with all of the science, ape charts, beast charts, and the veracity of no idea of proof, facts or, evidence of how it all got started and get to have no data and to insist that this big bang idea is absolutely the way and formation of how all of life begin.

Enthusiasm, religious enthusiasm

Logical commonsense fact:

To believe in an idea such as this with no proof, facts, or evidence and not even a complete idea of how it all got started and stand on it to continue one can only proceed by default on faith and belief in one's religious enthusiasm.

Evolution has tried to explain this away but the only way to continue believing that Evolution is true without any evidence is by faith alone.

In the nucleus of the Adam

What is the binding force of the nucleus gluons is the answer often times given by the evolutionary scientists gluons, there's no such thing as gluons? Well it is an invisible force that cannot be seen under the microscope with the naked eye or by anything anywhere so this is something else that the evolutionary scientists made up. Whenever they can't come out with an idea or something plausible I imagine they go in a corner and say let's make something up because it's never has to make any sense without being logical or reasonable or even scientific it doesn't have to even be science it only has to be a big enough lie. I mean because to allowed ideas that come out of the evolutionary science closet are just not basic logical commonsense, not intelligence and it shouldn't be allowed to have a part in the academia of the man producing. If you look at some of these men they are smart, intelligent individuals I mean top of the chart a student's some Harvard graduates men and women from some of the best schools in the country and the world Oxford, Cambridge just to get in the schools you have something like a 180 IQ. Its is just remarkable to me that these men of higher thinking capacity have not the ability to get this right and only people like work pass the masquerade can find the answer? Was this just too simply? To say this is what I believe this is my faith this is my religion?

Facts are: electrons of the atom whirl around the nucleus billions of times 1 billionth of a second and the nucleus of the atom consists of particles called neutrons and protons. Neutrons have positive charges and are therefore neutral but protons have positive charges:

one law of electricity is that like charges repel each other. Since all of the protons in the nucleus are positively charged, they should repel each other and scatter into space.

So why is it that this does not happen? The evolutionist cannot answer, the evolutionist won't even speculate they like to say gluons yes. But point me to one of those please! The evolutionist cannot. But, what holds them together? The God that made them that is what hold them together. But the evolutionist, whatever they are talking about appears to be invisible and anyway they have to have faith that is what they believe is working to keep the protons from repelling each other and scatter into outer space, because they can't see anything! In other words keeping our virtual bodies from exploding and scattering into space, keeping this planet from exploding and scatter into space yes it is the God that made them. We'll hopefully one day they will meet the God who made the given that they are very faithful believing people.

So what we have here is a religion

A lot of people have saved lives and that's okay. But in the United States it is unconstitutional to teach any religion in any federally funded institutions.

This is a religion now the people teaching this wasn't there to see the fresh fish changing to see any of the animals changing from one metamorphosis to another. There to see the half gill, half lungfish produced. There are no expedition that has complied any proof or facts, or evidence for all to see.

So in the absence of the PFE, by default evolutionist is not having a working model of their ideology to verify what has taken place what they believe on is faith.

Evolution is, because it is:

Biologically: impossible because there no facts to prove it.

Geographically impossible: because there are facts to disprove ideas.

Historically impossible: because history can only tell what is written.

Anthological and Genenemologcal impossible: because things change by comparison.

Mineralogical impossible: because there is only so much pressure in the ground.

Geological impossible: because there is only so much pressure in the world.

Homological and Anima-Logical impossible: because humans and animals can only be friends.

Un-provable and logically impossible: because religion cannot be proven it must be believed by faith

The above sciences and all we the people here on earth are the only sciences that we can put our hands on somewhere, and spend their lives studying the sciences and they know them within the categories above. There are other disciplines that also apply under the different sciences above and these are just some of them. When evolution has been put under the microscope when compared to the sciences listed above evolution comes up short, lacking, and assuming facts not presented in evidence. Why, the evolution is still having a complete theory on RNA. What is this is how the whole idea of evolution started. The evolutionist of today has taken Mr. Darwin's ideology past the point of his imagination Charles Darwin's went to his grave without figuring out this most important part of this ideology, I leave you with his most profound statements that he never saw in his lifetime come to fruition.

"Noting to the abundance of fossils, numerous transitional must be found to prove my theory."

In the absence of PFE to complete the ideals above by default evolution is nothing more than a religion.

Nothing more than a religion!

Evolution is impossibility!

SUMMARY

It is remarkable how the whole world has been taken in by the ideas of a few that are in the position of power or in a position of influence the science community has for the past 100 hundred and 50 years or so perpetrated this fraud all over the world to be something of the truth in stead evolution has only proven themselves to be religious enthusiast and they can be religious enthusiast if they want to but not at the world's expense and definitely not at the expense of the minds of the children growing up in the world today if one didn't know any better one would think that this was a conspiracy by these people in order to keep the truth from coming to the light of day. I make an accusation that walks like a duck, talks like walks about, quacks like about, evasive answers like about, doesn't accept the truth like all subjects like a dove, but maybe it's a fraud. This is a challenge to all of the people of the world who has had the mine to check out things that buzz around in their heads. People like the school board of different counties and states and cities in Atlanta and Colorado in a number of schools around the world that have stamped in all of their biology school books to illustrate that the information in their science books are not necessarily true but that they are a theory and not a fact. Some others have the same, evolution is a theory and not a fact people in different states around the country some teaching of evolution for a number of months and years for the own personal reasons some states have recanted for obvious reasons however this is not only buzzing around in the heads of a few people nowadays it is buzz around and head a lot of people nowadays because like myself there are lots of annoying questions being asked by everyday people like flies one's own vernacular. This is a bug that one can find out information on this is a bug that one can figure out and this is a bug that one can locate one only needs to look for the right fly squatter. This may take a little time because this type of bug won't just go away if you market it, this type of bug won't stop when you wave at it, or befriended it. How ever sometimes one has to get up set with oneself, one's, friends, one's family, and one's world! This should always happen sometimes but not very often but just like the trash if you don't take it out every day it will pile up so after a while after you complete your research and find your own facts, your own PFE, free all responses if you will, with an answer that you can contend with, one that you can apply logical, commonsense, question, point, answer, and fact, to. At this point if one can be dissatisfied enough to pull the wool from over one's own eyes and see the truth about evolution in the light of day under a microscope then, and only then can one place this bug out of existence because this bug never did.

Evolution is a fraud; it's nothing more than a religion!

By Elder J. Biggs Junior

Okay now mind you the all information that I have completed here can be viewed from the Internet thank you and have a nice day and God bless.

elderbiggs@yahoo.com